WIRELESS AD-HOC NETWORKS

Edited by **Hongbo Zhou**

Wireless Ad-Hoc Networks

http://dx.doi.org/10.5772/3438
Edited by Hongbo Zhou

Contributors

Hongbo Zhou, Takuya Yoshihiro, Paul Muhlethaler, Bartlomiej Blaszczyszyn, Tat Wing Chim, S. M. Yiu, Lucas C. K. Hui, Victor O. K. Li, Li Liu, Xianyue Li, Jiong Jin, Zigang Huang, Ming Liu, Marimuthu Palaniswami, Shiwen Mao, Yingsong Huang, Phillip Walsh, Yihan Li, Di Yuan, Vangelis Angelakis, Niki Gazoni

Published by InTech

Janeza Trdine 9, 51000 Rijeka, Croatia

Publishing Process Manager Sandra Bakic
Typesetting InTech DTP Team
Cover InTech Design Team

Additional hard copies can be obtained from orders@intechopen.com

Wireless Ad-Hoc Networks, Edited by Hongbo Zhou
p. cm.
ISBN 978-953-51-0896-2

Meet the editor

Dr. Hongbo Zhou is an associate professor in Department of Computer Science at Slippery Rock University of Pennsylvania. He obtained his Ph.D. degree on Computer Science and Engineering from Michigan State University in 2005. His research interests include: wireless networking, mobile ad-hoc networks, wireless sensor networks, and network security. Dr. Zhou has published several research papers in these areas in international conferences and journals. Dr. Zhou has served as program committee member for International Conference on Network and System Security, International Conference on Computer Communications and Networks, and International Conference on Parallel and Distributed Systems. He also served as reviewer for numerous international conferences, journals and workshops, including Communications of ACM, Wiley International Journal of Wireless Communications and Mobile Computing, Wiley International Journal of Security and Communication Networks, and ACM Transactions on Autonomous and Adaptive Systems. He has been invited to review research project proposals for Research Grants Council of Hong Kong.

Contents

Preface

A wireless ad-hoc network is a wireless network deployed without any infrastructure. In such a network, there is no access point or wireless router to forward messages among the computing devices. Instead, these devices depend on the ad-hoc mode of their wireless network interface cards to communicate with each other. If the nodes are within the transmission range of the wireless signal, they can send messages to each other directly. Otherwise, the nodes in between will forward the messages for them. Thus, each node is both an end system and a router simultaneously.

The wireless ad-hoc network can be divided into several subcategories. With a Mobile Ad-hoc Network (MANET), the nodes are free to move arbitrarily. Thus, the network topology keeps changing, and these nodes are constrained by the power supply and computation capability. While inFor some other scenarios like such as a wireless mesh network, some nodes stay fixed. Because these nodes have access to continuous power supply and the Internet, they have access to relatively less restrained resources. A special case of MANET is Vehicle Ad-hoc Network (VANET), in which the on-board computer is integrated with a transceiver and additional accessories and gadgets, which makes the communications between vehicles or between vehicle and roadside device possible. One feature of the VANET is that the nodes (vehicles) move along predefined trajectory (streets and roads). Another feature is that the lifetime of some wireless links is short-termed.

The wWireless ad-hoc networks has have been gaining its popularity due to several reasons: (1) there are an abundance of all sorts of mobile devices and sensors that support wireless connections; (2) without the necessity of infrastructure, it is fast, convenient, and economical for deployment; (3) it facilitates access to and sharing of digital information anywhere and anytime without human configuration or maintenance.

There are an array of applications of wireless ad-hoc networks in the place where there is no infrastructure or the infrastructure is knocked out, and a wireless ad-hoc network is the only option to share data and information between users. For instance, a temporary wireless network could be created to connect laptops and other devices during a field trip in a forest/desert/island, regardless ifno matter the users move around or stay put. With a MANET established among firefighters, police, and paramedics, they could share information to respond to an emergency more efficiently, which could mean saving a life. A similar scenario is that in the battlefield, soldiers, tanks, helicopters, and airplanes could use a MANET to share surveillance data in order to improve the precision and efficiency of attack and chance of survival. A wireless mesh network can be created among sensor nodes to collect and forward all sorts of data to a central station. A VANET is an indispensible component in a smart transportation system. Each car becomes the producer and consumer of traffic information, including but not limited to traffic light status, traffic jam, accidents, road work, weather, and parking lot vacancy. Other information, like the planned itinerary and the next turn/lane/exit to pick may also be shared with privacy concerns taken care ofconsidered. The collection of the data and information will travel along vehicle-to-vehicle links or the links between vehicles and road-side devices that are connected to the Internet. Eventually they will be fed into the driverless system to make commuting safer and more relaxing.

Due to the intrinsic characteristics of node instability, limited computation resources, bandwidth, and power supply, constant topology change, distributed operations, and lack of centralized management, it is challenging for the design and implementation of network applications and protocols for a wireless ad-hoc network. Generally speaking, although the idea and principles of design and protocols for a hardwired network can be referred to, they cannot be ported directly to a wireless ad-hoc network. Their design needs to be adapted or started from scratch to achieve efficiency or optimization under a new set of constraints.

Dr. Hongbo Zhou
Associate Professor,
Department of Computer Science,
Slippery Rock University, USA

Section 1

MAC Protocols for Wireless Ad-Hoc Networks

Chapter 1

Comparison of the Maximal Spatial Throughput of Aloha and CSMA in Wireless Ad-Hoc Networks

B. Blaszczyszyn, P. Mühlethaler and S. Banaouas

Additional information is available at the end of the chapter

http://dx.doi.org/10.5772/53264

1. Introduction

Multiple communication protocols are used to organize transmissions from several sources (network nodes) in such a way that scheduled transmissions are likely to be successful. Aloha is one of the most common examples of such a protocol. A major characteristic of Aloha is its great simplicity: the core concept consists in allowing each source to transmit a packet and back-off for some random time before the next transmission, independently of other sources. The main idea of the Carrier-Sense Multiple Access technique (CSMA) is to listen before sending a packet. CSMA is perhaps the most simple and popular access protocol that integrates some collision avoidance mechanism.

Simple classical models allow one to analyze Aloha and CSMA (see [1, 2]). They show that CSMA significantly outperforms Aloha as long as the maximum propagation delays between network nodes remain small compared to the packet transmission delays. However these models are not suitable for a wireless multihop network context, as they do not take into account the specificity of the radio propagation of the signal. Consequently, they cannot capture the spatial reuse effect (i.e., the possibility of simultaneous successful wireless transmissions) which is a fundamental property of multihop wireless communications.

Intuitively, it could be inferred that the collision avoidance embedded in CSMA should provide a greater spatial throughput than Aloha's purely random technique. Despite the large number of studies which evaluate Aloha and CSMA, to the authors' best knowledge there is no "fair" comparison of the spatial throughput of the two schemes in wireless multihop ad-hoc networks[1]. The aim of this paper is to carry out such a comparison and to quantify the gain in spatial throughput of CSMA over Aloha. We also study the effect of the various parameters on the performances. To do so, we model the geographic locations of network nodes by a planar Poisson point process and use the standard power-law path-loss

[1] [3] is the only similar study we know of but we explain in this paper why the comparison presented in [3] is not, in our opinion, "fair" according to us.

function of the Euclidean distance to model the mean attenuation of the signal power. Regarding radio channel conditions, we consider both standard Rayleigh and negligible fading. We use a SINR model in which each successful transmission requires that the receiver is covered by the transmitter with a minimum SINR.

For Aloha (both slotted and non-slotted), the above model lends itself to mathematical analysis as shown in [4, 5]. We adopt use and develop this approach and use simulations (which confirm the analytical results) to evaluate and optimize the performances of Aloha. The performance of the CSMA in the previous model is very complex thus we use simulations to study it.

The main contribution of this paper is the analysis and comparison of the performances of slotted, non-slotted Aloha and CSMA, all optimized to maximize the rate of successful transmissions, *under various radio propagation assumptions* (path-loss exponent, fading conditions). Our *main findings of this analysis are*:

- CSMA always outperforms slotted Aloha, which in turn outperforms non-slotted Aloha. In a moderate path-loss scenario (path-loss exponent equal to 4), without fading and the SINR level required for capture equal to 10, CSMA offers approximately a 2.4 times larger rate of successful transmissions than slotted Aloha and approximately a 3.2 times larger rate than non-slotted Aloha.
- The advantage of using CSMA is slightly reduced by increasing path-loss decay.
- This advantage is significantly reduced by the existence of fading since CSMA is much more sensitive to channel randomness than Aloha. In particular, for Rayleigh fading the above comparison of CSMA to slotted and non-slotted Aloha gives the ratios 1.7 and 2.3, respectively.
- The advantage of using CSMA increases with the SINR capture level.
- The above observations are valid when the transmissions are roughly scheduled to nearest neighbors and all the three MAC schemes are optimally tuned. This optimal tuning results in scheduling each node for transmission for about 8%, 6% and 4% of the time, for CSMA, slotted and non-slotted Aloha, respectively. These values do not depend on the network density, provided the nearest-neighbor receiver scheduling is used.
- The optimal tuning of CSMA is obtained by fixing the carrier-sensing power level (used to detect if the channel is idle) to about 8% of the useful signal power received at the nearest neighbor distance. This makes the transmissions successful with a high probability (from 0.8 to 0.95). Both smaller and larger values of the carrier-sensing threshold lead to essentially suboptimal performance of CSMA and sometimes even comparable to that of slotted Aloha. This might explain the apparent contradiction of our results to those of [3], which indicate similar performance of Aloha and CSMA.

This paper also *contributes to the development of the mathematical tools for Aloha by showing that the so-called spatial contention factor cf [6], appearing in the Laplace-transform characterization of the interference, is larger in non-slotted Aloha than in slotted Aloha under the same channel assumptions, by a factor that depends in a simple, explicit way only on the path-loss exponent; cf. Fact 3.2.* We also suggest the usage of the Bromwich contour inversion integral, developed in [7], to evaluate the coverage probability in the no-fading case; cf. Fact 3.6.

In this paper we will not take into account second order factors such as the back-off strategy in CSMA or guard intervals in slotted Aloha. We will briefly discuss these factors at the end of the paper to show that they cannot change the order of magnitude of the comparison between Aloha and CSMA.

The remaining part of this paper is organized as follows. In Section 1.1) we recall some previous studies of Aloha and CSMA. Section 2 introduces the model: distribution of nodes, channel and capture assumptions. It also describes in more detail the three MAC protocols studied in this paper. In Section 3 we present our analysis tools. Section 4 provides our findings regarding the performance of the MAC protocols considered. The conclusions are presented in Section 5.

1.1. Related work

Aloha and Time Division Multiple Access (TDMA) are the oldest multiple access protocol. Aloha, which is the "mother" of random protocols, was born in the early seventies, the seminal work describing Aloha [8] being published in 1970. Since that time it has become widespread in various implementations. The essential simplicity of Aloha also allows for simple analysis. A first, and now widely taught result regarding the ratio of successful transmissions (cf. e.g [2, 4.2]) was obtained assuming an aggregate, geometry-less process of transmissions following a temporal Poisson process, with some overlapping of two or more packet transmissions necessarily leading to a collision. In this model, the ratio of successful transmissions can reach $1/(2e) \approx 18\%$, when the scheme is optimized by appropriate tuning of the mean back-off time (intensity of the Poisson process). It was also shown, that this performance can be multiplied by 2 in *slotted-Aloha*, when all the nodes are synchronized and can send packets only at the beginning of some universal time slots.

Although Aloha was primarily designed to manage wireless networks, the lack of a geometric representation of node locations in the above model makes it unsuitable for wireless networks. To the authors' best knowledge, it is in the paper by Nelson and Kleinrock [9] that Aloha was first explicitly studied in a wireless context. The authors showed that under ideal circumstances with slotted Aloha the "expected fraction of terminals in the network that are engaged in successful traffic in any slot does not exceed 21%". Despite the very simple on-off wireless propagation model used in the paper, this result, as we will show, is surprisingly close to the results that can be obtained using more recent and more sophisticated, physical propagation and interference models (cf. [4, 6]) in the case of the fading-less channel model with the mean path-loss decay equal to 3.5. The key element of this latter approach is the explicit formula of the Laplace transform of the interference created by a Poisson pattern of nodes using Aloha. This analysis was recently extended to non-slotted Aloha in [5]. We adopt this approach and slightly extend it in the present paper.

In the widely referenced paper [10] another simplified propagation model was used to study local interactions of packet transmissions and the stability of spatial Aloha.

CSMA was studied in the 70s in [1] and in the 90s in articles such as [11]. In these studies, the spatial reuse is usually not considered. However a few articles such as [12–14] take it into account by modeling carrier sensing with a graph. Nodes within carrier sense are linked vertices in this graph. However this model only approximates the carrier-sensing and the capture effect. [15] uses the same model for CSMA to study the per-flow throughput in the network. Other simplified models of the carrier sensing and capture effect are proposed in [16, 17].

At the end of the 90s, an original and well referenced study tried to capture the behaviour of the IEEE 802.11 distributed medium access algorithm [18]. Although this study represented a step forward in the analysis of the IEEE 802.11 collision avoidance mechanism, [18] did not include an accurate model to capture interference. Thus the spatial throughput of IEEE 802.11 cannot be analyzed with this model. Although numerous papers are actually using models close to that of [18], they are all unable to compute the spatial throughput of IEEE 802.11.

In contrast to [18], [19] studies the behaviour of a CSMA network using a more realistic model for interference and for the capture of packets. However [19] cannot obtain closed formulas and [19] is actually a semi analytical model based on a Markov chain. Moreover this model can only handle a few dozen nodes. Thus it cannot easily compute average performance or investigate the effect of the network parameters. New models have recently appeared such as [20]. These models use the Matern hard core process to model the pattern of simultaneously transmitting nodes in a CSMA network. These models, which allow the spatial throughput to be evalutated, have many flaws. First, CSMA is not accurately modeled by the Matern hard core process. Secondly the interference is also only approximated. Lastly the formulas obtained in these models to obtain the throughput are complex and it is difficult to use them to optimize the protocol when we vary the network parameters. Despite the many papers trying to analyze the performance of CSMA with spatial reuse, we believe that none of these papers offers a method for precise and straightforward evaluation of the gain from using the collision avoidance mechanism of CSMA, in the same framework (infinite Poisson ad-hoc network) in which spatial Aloha can been analyzed. Thus, for this paper we chose to rely on simulations to estimate the performance of CSMA. We believe that, for our purpose, this approach offers a faster, more accurate method which is also easy to implement.

We also want to recall the original geometric approach, also by Nelson and Kleinrock, presented in [3]. Their seminal paper presents a comparison of the performance of Aloha and CSMA in the geometric setting with the simple on-off wireless propagation model. Such a comparison is also the goal of our present study which however uses a more realistic propagation and interference model (see above). Our conclusions appear to *differ* from those of [3], where the performance of CSMA is found comparable to Aloha. We show that CSMA, with an appropriately tuned sensing threshold, can essentially outperform Aloha. The reason for this difference is presumably not due to the different wireless channel models, but primarily because of a sub-optimal tuning of the CSMA in [3], consisting of too small a sensing range (taken to be equal to the transmission range). In that sense [3] does not provide a fair comparison of the spatial throughput of Aloha and CSMA whereas, we believe, our paper does. [21] also compares Aloha and CSMA but only in terms of outage probability; [21] does not derive the density of successful transmission.

2. Models

In this section we present the models, which will be used to evaluate and compare the performance of the CSMA and Aloha MAC schemes.

2.1. Distribution of nodes and channel model

The model that we use here was proposed in [4]; we call it *the Poisson Bipole model*. It assumes that the nodes of a Mobile Ad hoc NETwork (MANET) are distributed on the infinite plane according to a homogeneous, planar Poisson point process of intensity λ nodes per unit surface area (say per square meter). Each node of this network transmits a packet to its own dedicated receiver located at random within a distance r meters from it, which is *not* a part of the Poisson point process. In this paper we choose $r = a/\sqrt{\lambda}$, for some constant $a > 0$, i.e. of the *order of the mean distance to the nearest neighbor* in a Poisson point process of intensity λ. This choice mimics the nearest neighbor scenario. We also assume that every node has always a pending packet to send. We believe that this assumption represents the behaviour of a loaded network and allows us to compute the maximum throughput of the network in a multihop context.

Using the formalism of the theory of point processes, we will say that a snapshot of the MANET can be represented by an independently marked Poisson point process (P.p.p) $\widetilde{\Phi} = \{(X_i, y_i)\}$, where the *locations of nodes* $\Phi = \{X_i\}$ form a homogeneous P.p.p. on the plane, with an intensity of λ, and where the mark y_i denotes the location of the receiver for node X_i. We assume here that one receiver is associated with only one transmitter and that, given Φ, the vectors $\{X_i - y_i\}$ are i.i.d with $|X_i - y_i| = r$.

We assume that whenever node $X_i \in \Phi$ transmits a packet it emits a unit-power signal that is propagated and reaches any given location y on the plane with power equal to $F/l(|X_i - y|)$.

$$l(u) = (Au)^{\beta} \quad \text{for } A > 0 \text{ and } \beta > 2 \tag{1}$$

and $|\cdot|$ denotes the Euclidean distance on the plane. Regarding the distribution of the random variable F, called for simplicity fading, we will consider two cases:

- constant $F \equiv 1$, called *the no fading case*,
- exponential F of parameter 1; this corresponds to the *Rayleigh fading* in the channel.

2.2. Successful transmission

It is natural to assume that transmitter X_i *successfully transmits* a given packet of length B to its receiver y_i within the time interval $[u, u + B]$ if

$$\text{SIR} = \frac{F/l(|X_i - y_i|)}{\bar{I}} \geq T, \tag{2}$$

where T is some signal-to-interference (SIR) threshold and where $\bar{I}$ is the *average interference* suffered by the receiver y_i during this packet transmission interval

$$\bar{I} = \frac{1}{B} \int_{u}^{u+B} I(t)\, dt, \tag{3}$$

with

$$I(t) = \sum_{X_j \in \Phi, X_j \neq X_i} F_{j,y_i}/l(|X_j - y_i|)\mathbf{1}(X_j \text{ transmits at time } t). \tag{4}$$

Note that taking (2) as the successful transmission condition, we ignore any external noise. This is a reasonable assumption if the noise is significantly smaller than the interference power $\bar{I}$, which is the case in our setting.

2.3. MAC protocols

We will assume a saturated traffic model, i.e, that each node always has a packet to transmit to its receiver. The times at which any given node can transmit are decided by the Medium Access Protocol (MAC). In this paper we study three MAC protocols: CSMA, slotted Aloha and non-slotted Aloha.

2.3.1. CSMA

The basic rule of CSMA is very simple: *each node ready to transmit a packet listens first to the channel and transmits only if it finds the channel idle. Otherwise it waits for the channel to be idle and further postpones its transmission attempt for an additional random "back-off" time used to select a single node among the nodes blocked by the previous transmission.* We assume that this random "back-off" time is very small and we do not consider it in this study. This assumption is true if the ratio of the propagation plus detection time over the transmission time of the packet is very small. We discuss at the end of the article how to introduce corrective terms if propagation and detection times are not neglige.

To decide whether the channel is idle, the sender node computes the interference it receives I'. If $I' \leq \theta$, where θ is the *carrier-sense threshold* then the channel is "idle" otherwise it is busy. The carrier-sense threshold θ is the main, and in our model, the only parameter that will be tuned to maximize the density of successful transmissions and thus optimize the performance of the CSMA.

2.3.2. Slotted Aloha

Slotted Aloha supposes that all the network nodes are perfectly synchronized to some time slots (each of the length B of the packet, common for the whole network) and transmit packets according to the following rule: *each node, at each time slot independently tosses a coin with some bias* p which will be referred to as the *Aloha medium access probability* (Aloha MAP); *it sends the packet in this time slot if the outcome is heads and does not transmit otherwise.*

The Aloha MAP p is the main parameter to be tuned to optimize the access (see a precise description of the stationary space-time model in [5]).

2.3.3. Non-Slotted Aloha

In non-slotted Aloha all the network nodes independently, without synchronization, send packets (of the same duration B) and then back off for some exponential random time of mean ε. In a more formal description of this mechanism one assumes that, given a pattern of network nodes, the temporal patterns of their retransmission are independent (across the nodes) renewal processes with the generic inter-arrival time equal to $B + E$ where E is exponential (back-off) with mean ε. A precise description of this stationary space-time model, called the *Poisson-renewal model* of non-slotted Aloha can be found in [5]. The analysis of this Poisson-renewal model of Aloha is feasible although it does not lead to simple closed formulas. In [5] another model, called the *Poisson rain model*, of non-slotted Aloha has been proposed. The main difference with respect to the scenario considered above is that the nodes X_i and their receivers y_i are not fixed in time. Instead, we may think of these nodes as being "born" at some time T_i transmitting a packet during time B and "disappearing" immediately after. The joint space-time distribution of node locations and transmission instances $\Psi = \{(X_i, T_i)\}$ is modeled by a homogeneous Poisson p.p. in $2 + 1$ dimensions with intensity $\lambda_s = \lambda B/(\varepsilon + B)$. It might be theoretically argued that the Poisson rain model is a good approximation of the Poisson-renewal model when the density of nodes λ is large, and the time instances at which a given node retransmits are very sparse. Indeed, the performance of the Poisson-renewal model is shown in [5] to be very close to that of the Poisson rain model. Thus, in our analytical study of non-slotted Aloha we will use the results regarding the latter for simplicity, while in our simulations we use the former.

2.4. Network performance under a given MAC

MAC protocols are supposed to create some space-time patterns of active (transmitting) nodes that increase the chances of successful transmissions. MAC optimization consists in finding the right trade-off between the density of active nodes and the probability that the individual transmissions are successful.

The first step of the analysis of the above trade-off problem consists in evaluating how much a given MAC protocol contends to the channel; i.e., how many packets it attempts to send per node and per unit of time. In homogeneous models this can be captured by the *average fraction of time a typical node is authorized to transmit*. We will denote this metric by τ. By space-time homogeneity, $\tau\lambda$ is the spatial density of active nodes at any given time and thus τ can also be interpreted as the probability that a typical node of the MANET is active at a given time. In what follows we will call it the*channel occupation* parameter. The way it depends on the basic (tunable) MAC parameters will be explained later on.

A complete evaluation of the performance of a MAC protocol must establish the fraction of successful authorized transmissions. We will denote by p_c the *probability that a typical transmission by a typical node is successful* (given this node was authorized by the MAC to transmit). We call it the *coverage probability* for short. By (2) we have

$$p_c = \mathbf{P}^0\{\, F \geq l(r)T\bar{I}\,\}\,, \tag{5}$$

where the probability $\mathbf{P}^0$ corresponds to the distribution of the random variables for a typical node during its typical transmission; this can be formalized using the Palm theory for point processes. This expression will be the basis of our analytical evaluation of the coverage probability for both slotted and non-slotted Aloha in Section 3.3. We can notice that $\bar{I}$ is independent of F in (5) because our MAC schemes do not schedule transmissions according to the channel conditions at the receivers.

We define the optimal performance of a given MAC scheme as the situation where the mean number of successful transmissions per unit of surface and unit of time $\tau\lambda p_c$, called the *density of successful transmissions*, is maximized. For a given MANET density λ, this is equivalent to maximizing τp_c, which can be interpreted as the probability that a typical node is transmitting at a given time and this transmission is successful. Following this interpretation, we call τp_c *the mean throughput per node*. It will be analytically evaluated for both Aloha schemes and estimated by simulations for Aloha and CSMA MAC.

3. Analysis tools

3.1. Simulation scenarios

Our simulations are carried out in a square of 1000 m$\times$1000 m in which we generate a Poisson sample of MANET nodes with intensity $\lambda = 0.001$ nodes per square meter. For each MANET node we generate the location of its receiver uniformly on the circle of radius $r = a\sqrt{1000}$ m centered on this node. To avoid side effects, we consider a toroidal metric on this square. (Recall that, roughly speaking, rectangular torus is a rectangle whose opposite sites are "identified".) Given this metric we consider the distance dependent path-loss model (1) with some given path-loss exponent β and $A = 1$.

Typically β is larger than 2 and smaller than 6. 2 corresponds to free space propagation and 6 is for situations with a lot of obstacles and reflections. We will use the default value $\beta = 4$; in [22] the Walfishch-Ikegami model provides $\beta = 3.8$. However in some experiments, we try different values of β. For each pair of nodes we generate an independent copy of the exponential variable F in the case of Rayleigh fading or take $F \equiv 1$ in the no-fading case. Unless explicitly specified, our default value of the SIR threshold is $T = 10$ which is a widely used value.

For a given distribution of nodes we run the dynamic simulation for each of the three MAC schemes described in Section 2.3 with some particular choice of their main parameters: the carrier-sense threshold θ for CSMA, MAP p for slotted Aloha and mean back-off time for non-slotted Aloha. The packet transmission duration is always $B = 1$ unit of time. We count both the total number of packet transmissions and the number of successful transmissions during the simulation, whose total time is 4000 units of time. For CSMA, as already said, we ignore the time spent in back-off when a node, after having sensed the channel busy, finds the channel idle again before attempting another transmission. In the simulations we use very small back-off times to select the transmitting nodes and we neglect the time actually spent in these back-offs. Since each packet transmission takes $B = 1$ unit of time, dividing the number of transmissions by the simulation time and by the number of MANET nodes in the square, we obtain the one-network-sample estimators of, respectively, the average fraction τ of time a typical node is authorized to transmit and the mean throughput per node τp_c. We repeat the above experiment for 10 random choices of the network and take the empirical means of the above one-sample estimators. The error-bars in all simulation results correspond to a confidence interval of 95%. We use a home-made event-driven simulator specially dedicated to our simulation problem. This simulatot provides much faster simulation results than the ones we would obtain with on the shelf simulation tools.

3.2. Analytical results for Aloha MAC

The analytical results for Aloha are based on the (simple) calculation of the average fraction of time a typical node is authorized to transmit τ and a (more involved) calculation of the Laplace transform of the interference $\bar{I}$ that is the only variable of "unknown" distribution in the expression (5) of the coverage probability p_c.

3.2.1. Channel Occupation τ

It is straightforward to see that in slotted Aloha $\tau = p$. In the Poisson-renewal model of non-slotted Aloha $\tau = B/(B+\varepsilon)$; i.e., the ratio between the packet duration time and the mean inter-transmission time.

3.2.2. Interference Distribution

The basic observation allowing explicit analysis of the coverage probability for all our Poisson models of Aloha is that the distribution of the interference $\bar{I}$ under the Palm probability $\mathbf{P}^0$ in (5) corresponds to the distribution of the interference "seen" by an extra receiver added to the original MANET pattern (say at the origin) during an arbitrary period of time of length B (say in $[0, B]$). This is a consequence of Slivnyak's theorem.

Moreover, note that in the slotted Aloha MAC the interference $I(t) = I$ in (3) does not vary during the packet transmission and consequently $\bar{I} = I$. Furthermore, note that the pattern of nodes X_j, which emit at a given time slot and interfere with a given packet transmission (cf. expression (4)) is a Poisson p.p. of intensity $p\lambda$. This is a consequence of the independent MAC decisions of Aloha. The general expression of the Laplace transform $\mathcal{L}_I$ of I, which in this case is a Poisson shot-noise variable, is known explicitly. Here we recall the expressions for the special cases of interest.

Fact 3.1. *For the slotted Aloha model with path-loss function (1) and a general distribution of fading F with mean 1 we have:*

$$\mathcal{L}_I(\xi) = \exp\{-\lambda\tau A^{-2}\xi^{2/\beta}\kappa\}, \tag{6}$$

where $\kappa \geq 0$ is some constant depending only on the path-loss exponent and the distribution of the fading F. In particular

http://dx.doi.org/10.5772/53264

- $\kappa = \pi\Gamma(1-2/\beta)$ *in the no-fading scenario* $F \equiv 1$,
- $\kappa = 2\pi\Gamma(2/\beta)\Gamma(1-2/\beta)/\beta$ *with Rayleigh fading.*

The constant κ was evaluated in [4] for Rayleigh fading and in [6], for the no-fading scenario, where the name *spatial contention factor* was proposed for this constant. $\Gamma()$ is the classical gamma function.

Regarding the distribution of the averaged interference $\bar{I}$ in non-slotted Aloha, we have the following *new general* result.

Fact 3.2. *Assume the Poisson rain model of non-slotted Aloha with space-time intensity of packet transmissions* $\lambda_s = \lambda\tau$ *and the path-loss function (1). Assume a general distribution of fading F. Then the Laplace transform* $\mathcal{L}_{\bar{I}}(\xi)$ *of the averaged interference* $\bar{I}$ *is given by (6) with the spatial contention factor* $\kappa = \kappa_{non-slotted}$ *equal to*

$$\kappa_{non-slotted} = \frac{2\beta}{2+\beta}\kappa_{slotted}\,,$$

where $\kappa_{slotted}$ *is the spatial contention factor evaluated for slotted Aloha under the same channel assumptions.*

Proof. By (3), (4) and exchanging the order of integration and summation we express $\bar{I}$ in the following form

$$\bar{I} = \sum_{X_j\in\Phi, X_j\neq X_i} F_{j,y_i} H_j / l(|X_j - y_i|)\,,$$

where $H_j = \frac{1}{B}\int_u^{u+B} \mathbf{1}(X_j \text{ emits at time } t)\, dt$. In the Poisson rain model we have $\mathbf{1}(X_j \text{ emits at time } t) = \mathbf{1}(t - B \le T_j \le t)$, where T_j is the time at which X_j starts emitting. Integrating the previous function we obtain $H_j = h(T_j)$, where $h(s) = (B - |s|)^+/B$ and $t^+ = \max(0,t)$. Consequently, for the Poisson rain model represented by Poisson p.p. $\Psi = \{X_i, T_i\}$ (cf. Section 2.3.3) the averaged interference at the typical transmission receiver is equal in distribution to

$$\bar{I} \overset{distr.}{=} \sum_{X_j, T_j \in \Psi} F_j h(T_j)/l(|X_j|)\,,$$

where F_j are i.i.d. copies of the fading. Using the general expression for the Laplace transform of the Poisson shot-noise we obtain for the path-loss function (1)

$$\mathcal{L}_{\bar{I}}(\xi) = \exp\Big\{-2\pi\lambda_s \int_{-\infty}^{\infty}\int_0^{\infty} r\Big(1 - \mathcal{L}_F\big(\xi h(t)(Ar)^{-\beta}\big)\Big)\, dr\, dt\Big\}\,,$$

where $\mathcal{L}_F$ is the Laplace transform of F. Substituting $r := Ar(\xi h(t))^{-1/\beta}$ for a given fixed t in the inner integral we factorize the two integrals and obtain $\mathcal{L}_{\bar{I}}(\xi) = \exp\{-2\pi\lambda_s A^{-2}\xi^{2/\beta}\zeta\kappa\}$, where $\zeta = \int_{-\infty}^{\infty}(h(t))^{2/\beta}\, dt$ and $\kappa = \int_0^{\infty} r(1 - \mathcal{L}_F(r^{-\beta}))\, dr$. A direct calculation yields $\zeta = 2\beta/(2+\beta)$. This completes the proof. □

Remark 3.3. Regarding the ratio of the spatial contention parameters $\zeta = \zeta(\beta) = 2\beta/(2+\beta)$, that can be seen as the *cost of non-synchronization in Aloha* (cf Remark 3.5 below), note that in the free-space propagation model (where $\beta = 2$) it is equal to 1 (which means that the interference distribution, and so coverage probability, in slotted and non-slotted Aloha are the same). Moreover, $\zeta(\beta)$ increases with the path-loss exponent and asymptotically (for $\beta = \infty$) approaches the value 2. This was only conjectured in [5].

3.3. Coverage probability

Evaluating p_c from (5) is straightforward in the case of Rayleigh fading. Indeed, with F independent of $\bar{I}$ one has $\mathbf{P}^0\{F \geq l(r)T\bar{I}\} = \mathbf{E}^0[\exp\{-l(r)T\bar{I}\}] = \mathcal{L}_{\bar{I}}(l(r)T)$. By Facts 3.1 and 3.2 we have the following result.

Fact 3.4. *For the Aloha model with the path-loss function (1) and Rayleigh fading*

$$p_c = \exp\Big\{-\lambda\tau r^2 T^{2/\beta}\kappa\Big\}, \tag{7}$$

where

- $\kappa = 2\pi\Gamma(2/\beta)\Gamma(1-2/\beta)/\beta$ *for slotted Aloha and*
- $\kappa = 4\pi\Gamma(2/\beta)\Gamma(1-2/\beta)/(2+\beta)$ *for non-slotted Aloha.*

Remark 3.5. Note that due to our parametrization $r = a/\sqrt{\lambda}$ (which mimics the nearest-neighbor receiver model), the maximal mean throughput per node τp_c is achieved (in slotted or non-slotted Aloha with Rayleigh fading) for $\tau = \tau^* = \kappa^{-1}a^{-2}T^{-2/\beta}$ and it is equal to τ^*/e. In particular, by Fact 3.2, non-slotted Aloha achieves $\zeta = \zeta(\beta)$ times smaller maximal throughput than slotted Aloha, where ζ is the cost of non-synchronization in Aloha. The dependence of this cost on β is analyzed in Remark 3.3. Here, note only that the well-known result obtained for the simplified collision model with on-off path-loss function, and saying that slotted Aloha offers two times greater throughput than non-slotted Aloha (see [2, Section 4.2]) corresponds in our model to the infinite path-loss exponent; $\zeta(\infty) = 2$.

In the case of a general distribution of fading the evaluation of p_c from the Laplace transform $\mathcal{L}_{\bar{I}}$ is not so straightforward. Some integral formula, based on the Plancherel-Parseval theorem, can be used when F has a square integrable density. This approach however does not apply to the no-fading case $F \equiv 1$. Here we suggest another, numerical approach, based on the Bromwich contour inversion integral and developed in [7], which is particularly efficient in this case.

Fact 3.6. *For Aloha model with constant fading* $F \equiv 1$ *we have*

$$p_c = \frac{2\exp\{d/(Tl(r))\}}{\pi}\int_0^\infty \mathcal{R}\Big(\frac{1-\mathcal{L}_{\bar{I}}(d+iu)}{d+iu}\Big)\cos ut\,du\,, \tag{8}$$

where $d > 0$ is an arbitrary constant and $\mathcal{R}(z)$ denotes the real part of the complex number z.

As suggested in [7], the integral in (8) can be numerically evaluated using the trapezoidal rule and the Euler summation rule can be used to truncate the infinite series; the authors also explain how to set d in order to control the approximation error.

3.4. Carrier-sense scaling in CSMA

As mentioned above, a similar analysis of the performance of the CSMA scheme is not possible. If fact, neither the channel contention described by τ nor the distribution of $\bar{I}$ under $\mathbf{P}^0$ is easy to evaluate for this scheme. Here we want only to comment on some scaling results (with the node density λ) regarding the performance of CSMA.

Note that in the noiseless scenario (cf SIR condition (2)), with nearest-neighbor-like distance $r = a/\sqrt{\lambda}$ from transmitter to receiver, and the path-loss function (1) the SIR is invariant with respect to a homothetic transformation of the model; i.e., dilating all the distances by some factor, say γ. However, the received powers (as interference I' measured by the transmitters) scale like $\gamma^{-\beta}$. By the well known scaling property of the homogeneous Poisson p.p. [2], this implies that the *performance of the CSMA scheme* (values of τ and the distribution of $\bar{I}$) *in our network model is invariant with respect to the MANET density provided the carrier-sense threshold* θ *varies with* λ *as* $\theta = \theta(\lambda) = \theta(1)\lambda^{\beta/2}$.

In order to present our simulation results for CSMA in a scale-free manner, in Section 4.1 we plot the mean throughput τp_c if the function of the *modified carrier-sense* threshold is $\tilde{\theta} := \theta l(r)$ that can be seen as the power normalized by the *received* signal power (in contrast to θ that is normalized to the *emitted* signal power). This results in $\tilde{\theta} = \theta(1)\lambda^{\beta/2}(A\lambda^{-1/2})^{\beta} = \theta(1)A^{\beta}$ which does not depend on the density of the MANET.

Another way of presentating scale-free results is to express the carrier-sense threshold θ in terms of the *equivalent carrier-sense distance* R defined as the distance at which a unit of emitted power is attenuated to the value θ, i.e. satisfying $\theta = 1/l(R)$. In our path-loss model this relation makes $R = \theta^{-1/\beta}/A$. We will use this approach when comparing our optimal tuning of CSMA to that proposed in [3]; see Section 4.3.

4. MAC optimization and comparison results

In this section we present our findings regarding analysis and comparison of the performance of Aloha and CSMA.

4.1. MAC performance study

We study the mean throughput per node τp_c achieved by CSMA and Aloha under our default setting ($a = 1,\ \beta = 4, T = 10$) with and without fading, depending on the MAC parameters, which are carrier-sense threshold θ, MAP p and mean back-off time ε for, respectively, CSMA, slotted and non-slotted Aloha.

4.1.1. CSMA

Figure 1 presents the throughput τp_c achieved by CSMA versus the modified carrier-sense threshold $\tilde{\theta}$. Recall that $\tilde{\theta}$ is the carrier-sense threshold in ratio to the useful power at the received (at the distance $r = a/\sqrt{\lambda}$) [3]. This makes τp_c and $\tilde{\theta}$ independent of the MANET density; cf. Section 3.4. Our first observations are as follows.

[2] The dilation of a planar Poisson p.p. of intensity 1 by a factor $\gamma = \lambda^{-1/2}$ gives a Poisson p.p. of intensity λ.

[3] In other words, e.g. $\tilde{\theta} = 0.1$ means that the channel is considered by an emitter as idle if the total power sensed by it is at most 10% of the mean useful signal power received by its receiver.

Remark 4.1. In the absence of fading the maximum throughput of 0.068 (unit-size packets per unit of time and per node) is attained by CSMA when the carrier-sense threshold is fixed roughly at the level of $\tilde{\theta} = \tilde{\theta}^* = 0.08$. This optimal tuning of the carrier-sense threshold seems to be quite insensitive to fading. However, the optimal throughput is significantly reduced by fading. Rayleigh fading of mean 1, reduces the CSMA throughput to 63.2% compared with the no fading scenario.

This latter observation is easy to understand as the channel-sensing is done at the emitter and that fading at the receiver is independent of fading at the emitter.

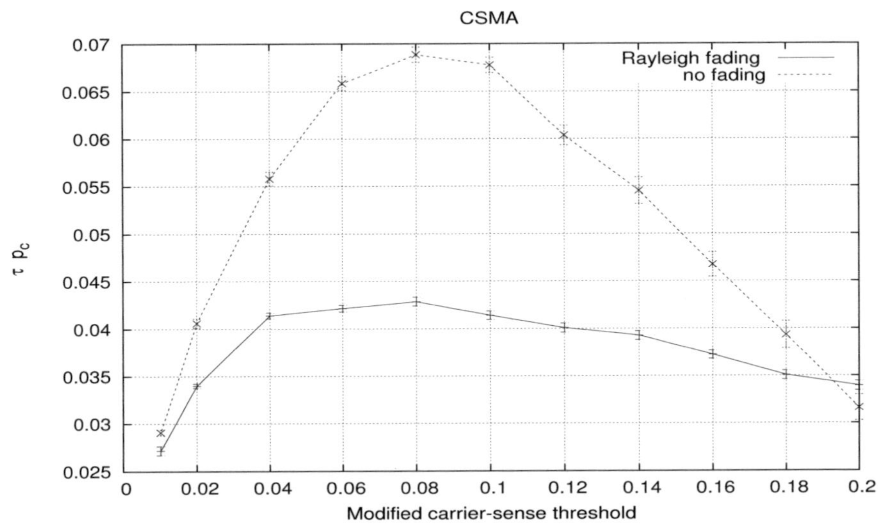

Figure 1. Mean throughput per node τp_c versus modified carrier-sense threshold $\tilde{\theta}$ in CSMA; Rayleigh fading and no-fading scenario.

4.1.2. Aloha

Figure 2 presents the throughput τp_c with and without fading achieved by slotted Aloha versus the channel occupation time τ, which in this model is equal to the the MAP parameter p. The results of non-slotted Aloha are presented in Figure 3 with $\tau = 1/(1+\varepsilon)$, where ε is the mean back-off time. The other parameters are as in the default setting. Here are our observations.

Remark 4.2. In the absence of fading the maximum throughput of 0.028 for the optimal MAP $p = p^* \approx 0.06$. As in CSMA, this optimal tuning seems to be quite insensitive to fading, which in the case of Rayleigh fading can be evaluated explicitly as $p = p^* = \kappa^{-1}T^{-2/\beta}$ (which gives $p^* = 0.064081$ in the default Rayleigh scenario). In contrast to CSMA, Rayleigh fading has a relatively small impact on the slotted Aloha throughput reducing it only to 92% of the throughput achieved in the no-fading scenario (in contrast to 63.2% in CSMA). Similar observations hold for non-slotted Aloha, which in the Rayleigh fading scenario achieves $\zeta = 2\beta/(2+\beta) = 1.5$ times smaller throughput than the slotted version.

4.2. Impact of model parameters

In Figures 4, 5, 6, 7, 8 and 9 we can study the dependence of the *maximal throughput* achievable by the MAC schemes (at their respective optimal tunings) as a function of the path-loss exponent β,

http://dx.doi.org/10.5772/53264

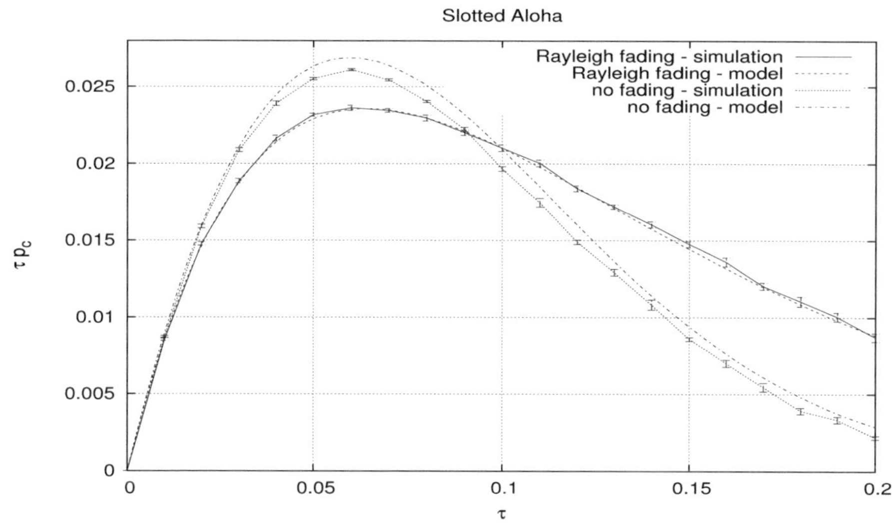

Figure 2. Mean throughput per node τp_c versus channel occupation $\tau = p$ in slotted Aloha; Rayleigh fading and no fading scenario.

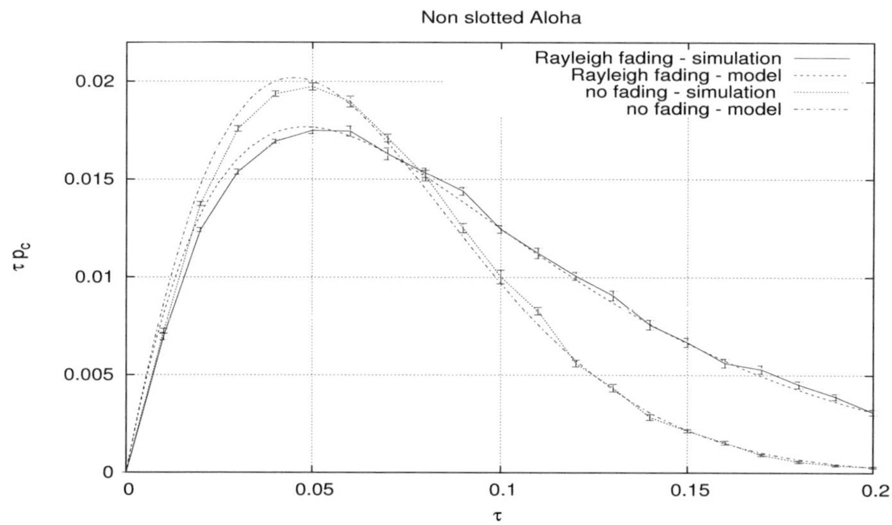

Figure 3. Mean throughput per node τp_c versus channel occupation time $\tau = 1/(1+\epsilon)$ for non-slotted Aloha; Rayleigh fading and no fading scenario.

SIR threshold T and relative distance to the receiver a (recall that $a = r\sqrt{\lambda}$). It is clear that CSMA significantly outperforms both Aloha protocols for all choices of parameters.

More detailed observations are as follows.

Remark 4.3. The higher path-loss exponent β is, the less advantage there is in using CSMA. When there is no fading, the increase of β from 3 to 6 reduces the gain in throughput of CSMA with respect to slotted Aloha from 2.6 to 2.1 and with respect to non-slotted Aloha from 3.5 to 3.2.

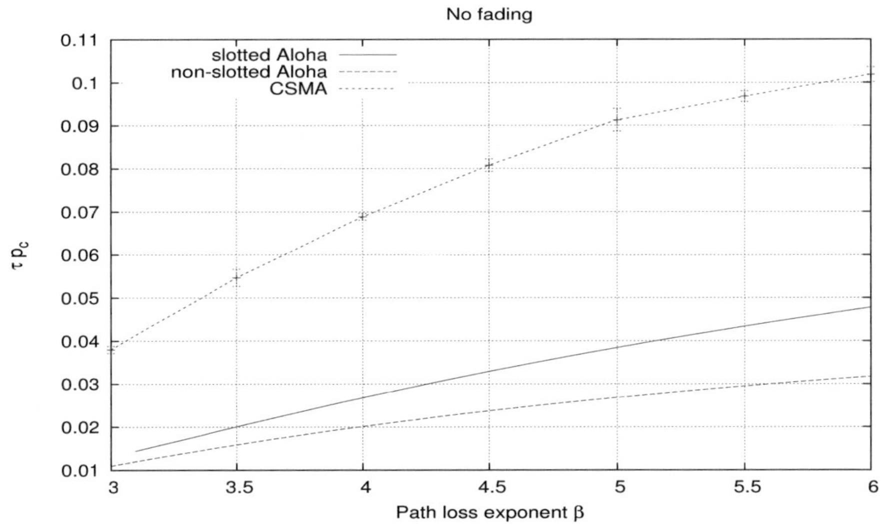

Figure 4. Maximal achievable mean throughput per node τp_c versus path-loss exponent β in the absence of fading.

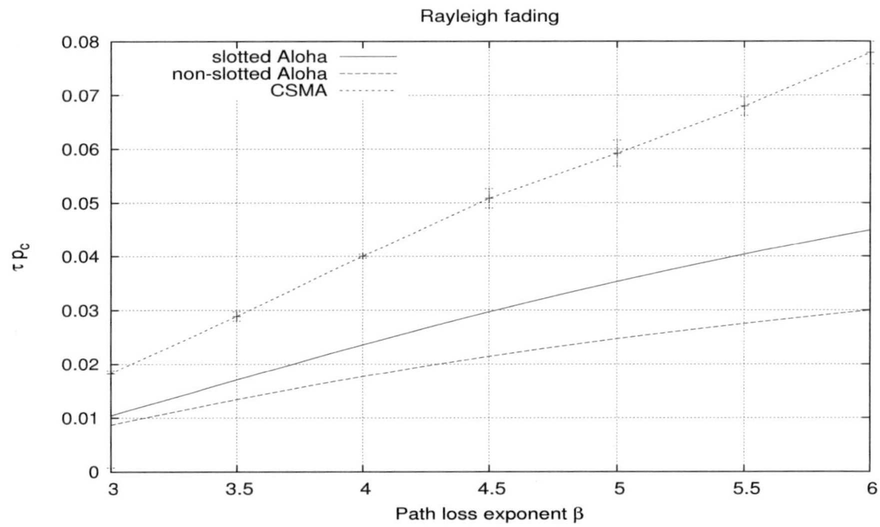

Figure 5. Maximal achievable mean throughput per node τp_c versus path-loss exponent β with Rayleigh fading.

We can also see in Figure 4, that in the absence of fading, slotted Aloha attains the expected fraction of 21% of terminals engaged in successful traffic, foreseen in the seminal paper [9], for SINR threshold $T = 10$ and a moderate path loss exponent slightly larger than $\beta = 3.5$.

Remark 4.4. The existence of fading (see Figure 5) further diminishes the advantage of CSMA. In particular, Rayleigh fading reduces the gain in throughput of CSMA with respect to slotted Aloha to about 1.7 and for non-slotted Aloha to a factor between 2.5 and 2.1 (depending on β).

Studying the impact of the SINR threshold T we observe the following, see Figures 6 and 7.

Remark 4.5. The higher T is (and hence the smaller bit-error rate sustainable in each packet), the greater is the advantage of using CSMA. In particular, when there is no fading and for $\beta = 4$, increasing

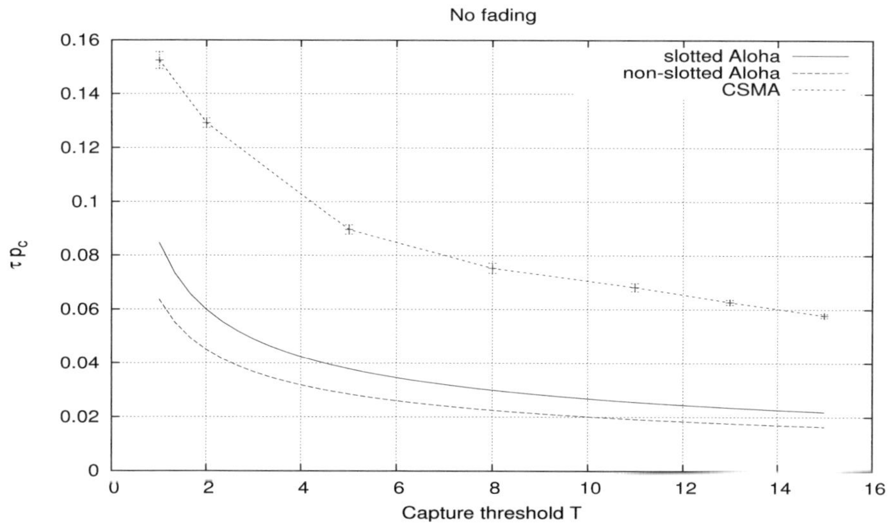

Figure 6. Maximal achievable mean throughput per node τp_c versus SIR threshold T in the absence of fading.

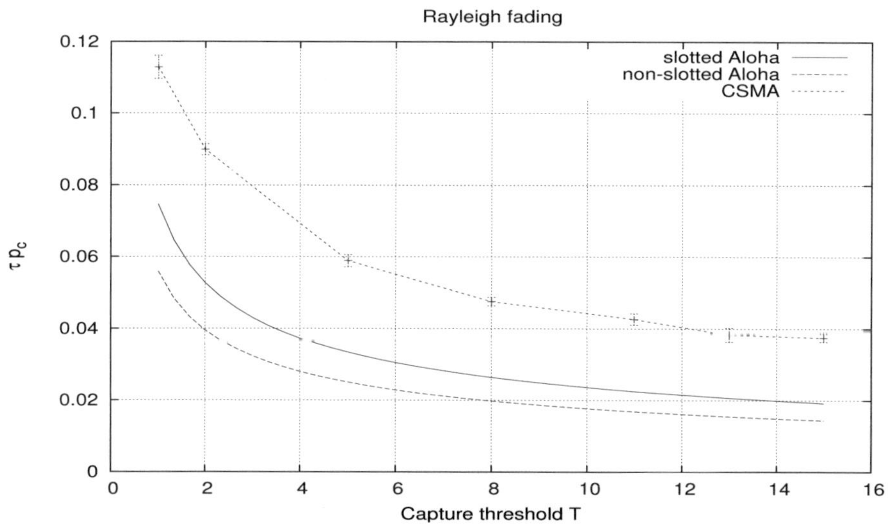

Figure 7. Maximal achievable mean throughput per node τp_c versus SIR threshold T. Rayleigh fading.

T from 1 to 11 results in the increase in the gain in throughput of CSMA with respect to slotted Aloha from 2.4 to 3.5 and this latter ratio remains stable for T larger than 11. For a similar comparison of CSMA to non-slotted Aloha the gains are from 1.8 to 2.6. In the case of Rayleigh fading the analogous gain factors of CSMA are, respectively, from 1.8 to 2.4 with respect to slotted Aloha and from 1.4 to 1.8 with respect to non-slotted Aloha.

Finally we study the impact of the relative distance to the receiver a (in ratio to the mean distance to the nearest neighbor in the network). Figures 8, and 9 show clearly that this distance should be kept as small as possible without disconnecting the network.

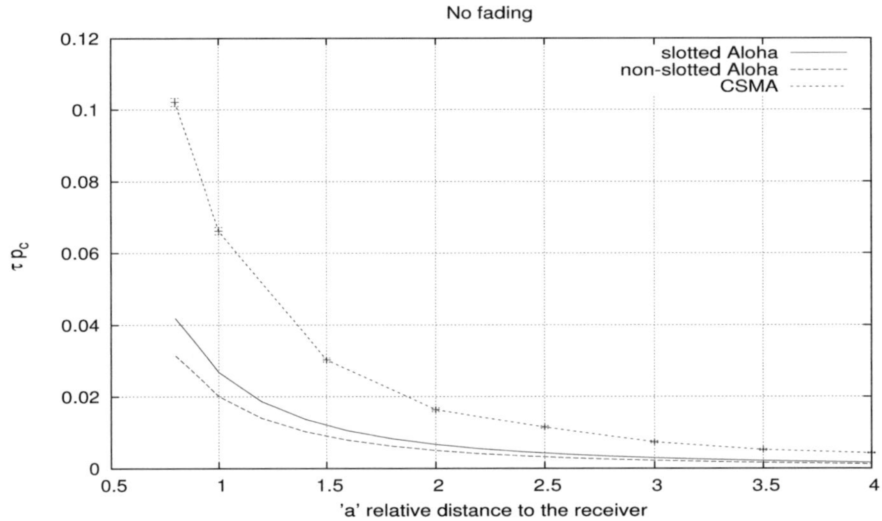

Figure 8. Maximal achievable mean throughput per node τp_c versus a the distance transmitter receiver. No fading.

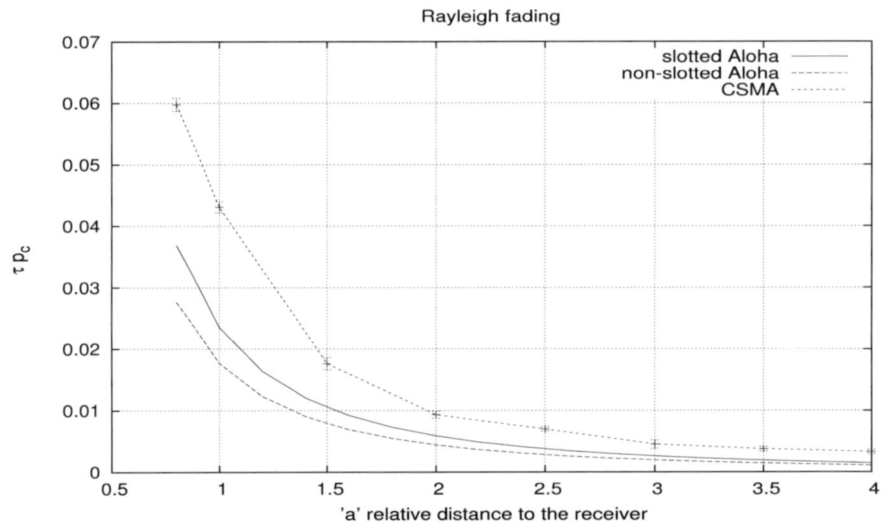

Figure 9. Maximal achievable mean throughput per node τp_c versus a the distance transmitter receiver. Rayleigh fading.

4.3. Optimal tuning of Aloha and CSMA

For Aloha the optimal tuning of τ can be obtained analytically from (7) whereas the optimal tuning of CSMA is obtained by simulation. In Figure 10, we present the optimal values of τ versus β with Rayleigh fading for both Aloha and CSMA.

Remark 4.6. We observe that the more sophisticated the MAC scheme is, the more it can content to the channel when the MAC is tuned optimally. Additionaly the more sophisticated MACs also exhibit higher capture probabilities. In particular our simulations show the that this probability is close to 1 (between 0.8 and 0.95) for CSMA.

A practical conclusion that can be drawn from these observations is that *the carrier-sense threshold in CSMA should be chosen at the largest possible value at which the allowed transmissions are almost always successful.*

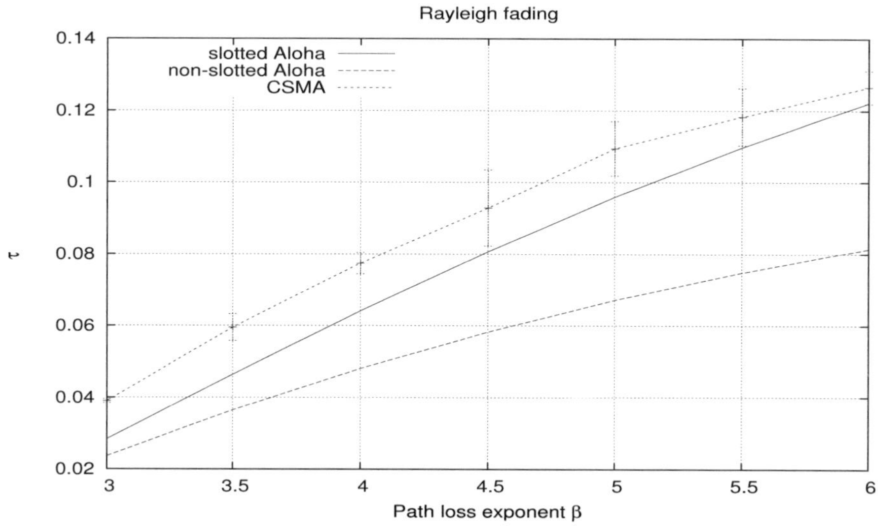

Figure 10. Optimal value of τ for Aloha and CSMA versus β with Rayleigh fading

4.4. Nelson & Kleinrock's model of CSMA revisited

Remark 4.6 might explain why the significant superiority of CSMA with respect to Aloha was not observed in [3]. Let us be more precise and revisit this model.

The simple propagation model in [3] assumes a fixed transmission range R and *the same* carrier-sense range. In other words, any two successfully communicating nodes need to be within distance R from each other and no other transmission should occur in the distance R from the receiver.

In this model, an *ideal medium access scheme* suggested in [3] should be able to choose from the given pattern of nodes centers for a maximal number of hard (non-intersecting) disks of radius R. The asymptotic analysis of the performance of such an ideal scheme is done in [3] assuming an increasing density of nodes λ. Namely, if this density is large, then the optimal scheme should be able to chose the pattern of nodes close to the hexagonal packing, known to obtain the densest packing of hard disks of radius R. Such a packing attains the fraction of 0.90689 of the plane covered by the union of disks. Consequently, since there is no disk overlapping, it would choose the fraction

$$\tau_{ideal} = \frac{0.90689 \times \text{network area}}{\# \text{ of nodes} \times \text{exclusion disk surface area}} = \frac{0.90689}{\lambda \pi R^2}$$

of the nodes of the network, whose density is λ nodes per unit of surface. This expression can be interpreted as the contention parameter of this ideal medium access scheme, which explains our notation. Since all transmissions allowed by this scheme are successful, we have $p_c = 1$ for it and the achieved throughput per node is $0.90689/(\lambda\pi R^2)$.

Regarding CSMA, the simple propagation model with transmission range equal to the carrier-sense range, assumed in [3], corresponds to a choice of nodes such that any selected node is not covered by the transmission range of any other selected node. This task is equivalent to the packing of hard disks of radius $R/2$. For some reason, that is partially explained in that paper, a slightly larger radius $1.2881R/2$ is taken. Similar to the ideal scheme, asymptotic analysis of the hexagonal pattern, gives the contention parameter of this CSMA scheme equal to

$$\tau_{CSMA} = \frac{2.214}{\lambda \pi R^2}.$$

Moreover, assuming that each authorized node chooses its receiver uniformly within the transmission range R, and calculating the fraction of the area within this range that is not covered by any other disk (no collision), the successful transmission probability is calculated as $p_c = 0.2034$. Consequently the throughput achieved by this CSMA is $\tau p_c = 0.4504/(\lambda \pi R^2)$.

Note that apparently the *sub-optimal assumption of the carrier-sense range equal to the transmission range in the above model of CSMA leads to a relatively small successful transmission probability* $p_c = 0.2034$, close to that obtained by Aloha, which explains why there is no essential difference between the performance of these two schemes. Our optimally tuned CSMA model seems to be closer to the ideal scheme of [3], at least because the probability of successful transmission is much closer to 1.

Let us now try to compare the performance of our optimal CSMA and the two schemes of [3]. This is not straightforward, since unlike ours, the results of [3] scale in $1/\lambda$ are only valid asymptotically, when $\lambda \to \infty$ (due to the hexagonal approximation of the perfect packing). However, note that in the model of [3], the expression $N = \lambda \pi R^2$ corresponds to the expected number of nodes within the area contended (blocked) by one given authorized transmission. Consequently the constants $\rho = 0.90689 = \tau p_c N$ and $\rho = 0.4504 = \tau p_c N$ can be interpreted, respectively in the two models, as the expected number of successful transmissions per set of nodes contended (blocked) by one given authorized transmission. This kind of spatial efficiency can be evaluated in our model using the notion of the equivalent carrier-sense distance $R = \theta^{-1/\beta}/A = r\tilde{\theta}^{-1/\beta}$ introduced in Section 3.4. Taking $N = \lambda \pi R^2$ with R calculated as such we obtain for our CSMA $\rho = \tau p_c N = \tau p_c \lambda \pi r^2 \tilde{\theta}^{-2/\beta}$. For the optimally tuned CSMA in the standard scenario $a = 1, T = 10, \beta = 4$ without fading we have $\rho = 0.07\pi 0.08^{-1/2} = 0.77750$ of successful transmissions per set of nodes contended (blocked) due to one given authorized transmission. This is a much better performance than $\rho = 0.4504$ for CSMA of [3], and in fact closer to $\rho = 0.90689$ achieved by the ideal scheme of [3].

4.5. Corrective terms

In this article we have not considered the effect of the back-off for CSMA, and for slotted Aloha we have ignored the guard times to avoid overlapping of the slots. In this sub-section we briefly study the effects of these parameters on performance. Let us call δ the ratio of maximum propagation time plus detection time in the network over the packet transmission time. Back-off in CSMA leads to wasting time and to collisions for nodes starting their transmissions within the same mini-slot of size $\delta\times$ packet transmission time. We know that the reduction of the throughput for CSMA is $\frac{1}{1+\sqrt{2\delta}}$ when the back-off is properly tuned, see [2] chapter 4. For slotted Aloha and with the same assumptions, the guard times lead to a reduction of approximately $\frac{1}{1+\delta}$. Thus for $\delta = 0.05, 0.02$ and 0.01 we obtain a throughput reduction of respectively $0.76, 0.83, 0.87$ for CSMA and $0.95, 0.98, 0.99$ for slotted Aloha. Thus, the throughput reduction is greater for CSMA than for slotted Aloha but these corrective terms do not change our main observation which gives a notably higher throughput to CSMA.

5. Conclusions

In this paper we compare slotted and non-slotted Aloha with CSMA in a Poisson ad-hoc network setting with SINR-based capture condition. We assume the usual power-law path-loss function and both Rayleigh and no-fading scenarios. To obtain a fair comparison between these protocols, their parameters are tuned to achieve the maximum successful transmission rates. Our analysis shows that CSMA always outperforms both slotted and non-slotted Aloha. However the gain obtained when using CSMA is slightly reduced by increasing path loss and more significantly by the existence of fading. We also show how to tune the carrier-sense threshold in CSMA so as to obtain its optimal throughput for an arbitrary network density. Our models concur with those of [3] even though some results may appear, at first glance, to be somewhat contradictory, because in [3] CSMA is not optimized.

Author details

B. Blaszczyszyn[1,2], P. Mühlethaler[1] and S. Banaouas

1 INRIA Paris-Rocquencourt, France
2 ENS Paris, France

References

[1] L. Kleinrock and F. A. Tobagi. Packet switching in radio channels: Part I-Carrier-sense multiple-access modes and their throughput-delay characteristics. *IEEE Trans. Commun., vol. COM-23*, pages 1400–1416, 1975.

[2] D. Bertsekas and R. Gallager. *Data Networks*. Prentice-Hall, Englewood Cliffs, 2001.

[3] R. Nelson and L. Kleinrock. Maximum probability of successful transmission in a random planar packet radio network. In *Proc. of IEEE INFOCOM*, San Diego, April 1983.

[4] F. Baccelli, B. Blaszczyszyn, and P. Mühlethaler. An Aloha Protocol for Multihop Mobile Wireless Networks. In *Proceedings of the Allerton Conference, University of Illinois, Urbana Champaign*, November 2003. also in *IEEE Transactions on Information Theory*, 52(2):421–436, 2006.

[5] B. Błaszczyszyn and P. Mühlethaler. Stochastic analysis of non-slotted Aloha in wireless ad-hoc networks. In *Proc. of IEEE INFOCOM*, San Diego, CA, 2010.

[6] M. Haenggi. Outage, local throughput, and capacity of random wireless networks. *IEEE Trans. Wireless Comm.*, 8:4350–4359, 2009.

[7] J. Abate and W. Whitt. Numerical inversion of laplace transforms of probability distributions. *ORSA Journal on Computing*, 7(1):38–43, 1995.

[8] N. Abramson. The Aloha system - another alternative for computer communication. In *Proc. of AFIPS*, pages 295–298, 1970.

[9] R. Nelson and L. Kleinrock. The spatial capacity of a slotted Aloha multi-hop packet radio network with capture. *IEEE Trans. Comm.*, 32:684–694, 1984.

[10] C. Bordenave, S. Foss, and V. Shneer. A random multiple access protocol with spatial interactions. In *Proc. of WiOpt*. IEEE, Limassol, Cyprus, 2007.

[11] J. Kim and Leu. Capture Effects of Wireless CSMA/CA/Protocols in Rayleigh and Shadow Fading Channels. *IEEE Transactions on Vehicular Technology*, 48:1277–1286, 1999.

[12] R. Boorstyn, A. Kershenbaum, B. Maglaris, and V. Sahin. Throughput analysis in multihop csma packet networks. In *IEEE Transactions on Communications vol. 35, no. 3*, pages 267–274, 1987.

[13] José M. Brazio and Fouad A. Tobagi. Theoretical results in throughput analysis of multihop packet radio networks. In *In Proceedings of ICC*, pages 448–455, 1984.

[14] M. Garetto, T. Salonidis, and E. Knightly. Modeling per-flow throughput and capturing starvation in csma multi-hop wireless networks. In *IEEE/ACM Transactions on Networking*, pages 864–877, August 2008.

[15] Michele Garetto, Theodoros Salonidis, and Edward W. Knightly. Modeling per-flow throughput and capturing starvation in CSMA multi-hop wireless networks. In *In Proc. of IEEE Infocom*, Barcelona, SPAIN, 2006.

[16] K. Medepalli and F.A. Tobagi. Towards performance modeling of IEEE 802.11 based wireless networks: A unified framework and its applications. In *Proc. of IEEE INFOCOM*, Barcelona, SPAIN, 2006.

[17] Yu Wang and J.J. Garcia Luna-Aceves. Modeling of collision avoidance protocols in single-channel multihop wireless networks. In *Wireless Networks, Volume 10 Issue 5*, 2004.

[18] G. Bianchi. Performance analysis of the ieee 802.11 distributed coordination function. In *IEEE Journal on Selected Areas in Communications*, volume 18 No.3, pages 535–547, 2000.

[19] Lili Qiu, Yin Zhang, Feng Wang, Mi Kyung Han, and Ratul Mahajan. A general model of wireless interference. In *MOBICOM*, pages 171–182, 2007.

[20] Huu Quynh Nguyen, François Baccelli, and Daniel Kofman. A stochastic geometry analysis of dense ieee 802.11 networks. In *INFOCOM*, pages 1199–1207, Anchorage, Alaska, 2007.

[21] Mariam Kayna and Nihar Jindal. Performance of Aloha and CSMA in spatially distributed networks. In *ICC*, pages 1108–1112, 2008.

[22] European Commission. *COST 231. Evolution of land mobile radio (including personnal) communications. Final report. Information, Technologies and Sciences*. Springer, 1999.

Chapter 2

A Distributed Polling Service-Based Medium Access Control Protocol: Prototyping and Experimental Validation

Yingsong Huang, Philip A. Walsh, Shiwen Mao and Yihan Li

Additional information is available at the end of the chapter

http://dx.doi.org/10.5772/53772

1. Introduction

Mobile ad hoc networks and its variations such as wireless mesh networks and wireless LANs (WLAN) have become the ubiquitous connectivity solution in public as well as residential access networks, due to their cost efficiency, reliability and flexibility of deployment and operation. The rapidly proliferation of such wireless access networks are greatly advanced by the distributed multiple access control (MAC) protocols, which is based on random access techniques such as ALOHA, slotted ALOHA, carrier sense multiple access (CSMA) and CSMA with collision avoidance (CSMA/CA). The most important standards for these applications are the protocols in the IEEE 802.11 [1] series, which are widely used as the solution for the "last mile" access problem and become a *de facto* standard for various wireless access networks. The IEEE 802.11 protocol family defines physical layer (PHY) and medium access control (MAC) functions for wireless communication in the ISM bands of 2.4GHz and 5GHz. There are various amendments for the standard 802.11, such as 802.11a/b/g/e/n and the currently working draft of 802.11ac. Most of these amendments focus on the enhancement in PHY, which provides higher link capacity. For example, 802.11g adopts OFDM to leverage the data rate up to 54Mbps in 2.4GHz band. 802.11n [2] further improves the previous standards by adding multiple-input multiple-output (MIMO) antennas and the link capacity is boosted up to 600 Mbps. Although various PHY techniques are added to improve the link capacity, the MAC's they are based on almost remains same, which is based on CSMA/CA.

1.1. IEEE 802.11 MAC's

IEEE 802.11 employs Distributed Coordination Function (DCF) based on a CSMA/CA MAC protocol with binary exponential backoff algorithm. An optional access method by using Point Coordination Function (PCF) is also defined in IEEE 802.11 standard. PCF is designed for the operation in the Access Point (AP) mode, which allows the AP to poll nodes and transmit beacon frames to the nodes. PCF is only supported with AP thus limits its application in ad hoc networks mode. Due to its limitation, only very few products support

PCF in reality. We focus our discussion on DCF in the book chapter, which supports both infrastructure mode and ad hoc network mode, and is fully implemented in all commercial WLAN devices.

DCF follows the CSMA/CA techniques with random backoff algorithms. It also defines an optional handshake of Request-To-Send and Clear-To-Send (RTS/CTS) to reduce frame collisions introduced by the hidden node problem. In a single hop network, a source station with data to transmit senses the channel. After the channel is sensed idle for the period of DCF inter-frame space (DIFS), the station starts the random backoff by decreasing its contention window (CW). At the beginning of the backoff procedure, the source station shall generate a random backoff period for additional deferral before transmitting, unless the backoff timer already contain a nonzero value. The backoff period is randomly generated in $[0, CW - 1]$. CW takes an initial value of CW_{min}, and is doubled after each collision or unsuccessful transmission, until it reaches CW_{max}. After a successful transmission, CW will be reset to CW_{min}. If the channel is busy before CW reaches 0, CW will be frozen. The transmission will then be deferred and the station restarts to seek the idle status of DIFS interval of the channel. If the channel keeps idle when CW reaches 0, an RTS frame with the transmission duration will be transmitted. When the destination station receives the RTS, it may return a CTS frame to confirm that it is ready for receiving data. The CTS frame also contains the transmission duration, which may contains the DATA frame durations and allow other stations set up their Network Allocation Vector (NAVs) for virtual carrier sensing. The neighbor stations then go to sleep mode and come back to sense the channel after their NAV expires. After receiving the CTS, the source station will transmit one DATA frame to the destination. All other stations will keep silent and wait for the NAV to expire. When the DATA frame is received, an ACK frame is issued by the destination to acknowledge the successfully received frame after the period of short inter-frame space (SIFS). If the source station does not receive the ACK frame within a specified ACK timeout interval, the backoff procedure will be performed at the source station to defer the transmissions. The lost or corrupted frame will be retransmitted at a later time. A typical DCF process and backoff algorithm are shown on Fig. 1 and Fig. 2.

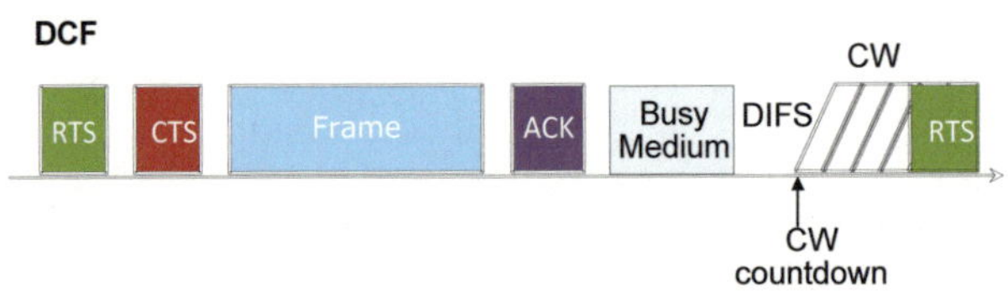

Figure 1. Illustration of DCF.

The original DCF scheme does not differentiate the traffic of different network services. It treats the high priority traffic and low priority traffic equally, which is not capable of satisfying the Quality of Service (QoS) request from the applications, such as voice over wireless LAN and streaming multimedia. To enhance the support of QoS, IEEE 802.11e [3] extends the MAC's by introducing Hybrid Coordination Function (HCF), which divides traffic into different classes and guarantees a QoS to each class. In the service differentiation, traffic in the same class competes the channel fairly like "best effort" transmission scheme, while traffic from different classes obtains different level of service. Although this service differentiation idea does not guarantee hard QoS, such as delay and loss rate, it provides a better response to the QoS requirements for different classes of services. The main techniques

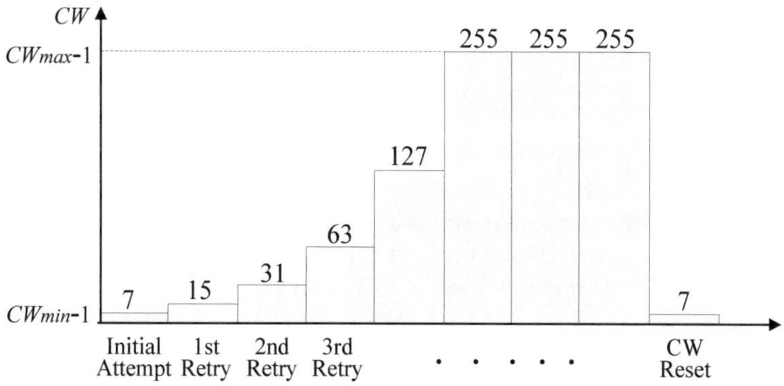

Figure 2. The *CW* backoff procedure.

used in the service differentiation of 802.11e include Enhanced Distributed Channel Access (EDCA) and HCF Controlled Channel Access (HCCA) components. The former is for contention-based channel access by extending DCF, while the latter is for contention-free transfer by extending PCF.

EDCF classifies the medium access according to the priority of access classes (AC). Intuitively, it can be noticed that the length of DIFS in DCF controls the priority of transmitting RTS frame. In EDCF, an arbitration inter-frame space (AIFS) is defined to specify the minimum number of slots for which the stations in the AC should sense the channel to be free before attempting transmission. The station in higher priority AC is assigned shorter length of AIFS and the *CW* will countdown earlier than lower priority nodes, hence will have a higher success probability. Further, different random backoff window size settings CW_{min} and CW_{max} can be used for different ACs. High priority traffic has a higher transmission chance than the low priority traffic by assigning smaller CW_{min} and CW_{max}.

Polling has been adopted in wireless MAC protocols. For example, the master-driven architecture of Bluetooth piconets provides an ideal setting for applying polling-based scheduling. Polling is adopted in Bluetooth piconets, but the actual scheduling policy has not been prescribed in the current standard [4]. The polling mechanism has been also incorporated in the HCCA. The hybrid coordinator (HC) polls QoS enhanced stations (QSTA), to assign them transmission opportunities (TXOP). A TXOP is a bounded time interval in which a QSTA is allowed to transmit one or more frames. Again, the specific scheduling policy has not been specified.

Recently, the reverse direction protocol has been suggested for IEEE 802.11n to support higher speed and higher throughput [5]. This technique gives an opportunity for a receiver to transmit data to a sender during the sender's TXOP, which is suitable for the highly asymmetrical traffic network applications, such as FTP and HTTP. Since the NAV duration may be changed in CTS to support the "bidirectional" TXOP, more complex schemes are needed to handle hidden nodes problems.

IEEE 802.11 MAC, although widely used in WLANs, they are well-known for their considerable control overhead, which could consume as much as 40% of the nominal link capacity [6]. For example, the maximum achievable throughput for IEEE 802.11a is 24.7 Mbps, which is about 45.7% of the normianl link capacity. The problem gets even worse in the multi-hop scenario, due to carrier sensing and spatial reuse issues [7]. The compelling demands to support high definition videos, online games, and other real-time applications bring new challenges to the usage efficiency of the link capacity of existing WLANs and stress the new design of more effecient wireless MAC's.

1.2. Polling Service-Based MAC

We presented three polling service-based MAC protocols, termed PSMACs in our prior work [8, 9], which can amortize the control overhead of medium contention/resolution over multiple back-to-back frame transmissions, thus achieving high efficiency in medium access control. The gated service based PSMACs are analyzed and compared with p-Persistent CSMA, which closely approximates the standard IEEE 802.11 DCF [10]. Considerable gains on throughput, delay, energy consumption, and fairness performance are observed in the analysis and simulation studies [9].

There are two fundamental differences between the proposed PSMACs and the existing polling approaches in IEEE 802.11 series. First, the schemes adopted in Bluetooth and HCCA are centralized ones, where a master or base station polls other stations. They are designed for relatively simple network topologies (e.g., a piconet with one master and seven slaves [4] or a single-hop WLAN). However, there may be no such master/base station in distributed wireless networks. These centralized approaches are quite different from the random access and fully distributed approach taken in PSMAC. Second, even for single-hop networks, the specific scheduling policy is not specified in either Bluetooth or IEEE 802.11 MACs. More importantly, there is a need of both theoretical and experimental study to underpin the scheduling techniques to be adopted in both standards.

In this book chapter, we introduce PSMACs protocols and prototype the PSMACs in a real wireless networking environment [11]. Generally, testbeds can provide useful insights that computer-based simulations cannot offer, since they capture the complex real-world radio propagation effects as well as distributed network dynamics, which are often greatly simplified in simulation and theoretical studies to make the problem manageable. By prototyping PSMACs, we can not only evaluate the MAC protocols under realistic wireless channels and verify our prior theoretical and simulation studies, but also identify new practical constraints and problems.

Two main contributions are made in this work. First, we implement the PSMACs on the GNU Radio [12] and Universal Software Radio Peripheral (USRP) [13] platform. We integrate the key functions of 802.11 DCF and the gated service policy in the implementation, such as gated service scheduling, CSMA/CA, virtual carrier sensing, RTS/CTS handshake, automatic repeat request (ARQ), random backoff mechanism, and distributed clock synchronization using IEEE 1588. Second, we conduct extensive experiments with various traffic types and traffic patterns, to evaluate the real system performance of the PSMAC testbed in both infrastructure mode and ad hoc mode. The experimental results demonstrate the significant improvements that PSMAC can achieve on throughput, delay and fairness, and also validate the theoretic analysis and simulation studies in the prior work [9].

The remainder of this chapter is organized as follows. We first review PSMAC in Section 2. We then provide the system overview in Section 3 and discuss implementation details in Section 4. Our experimental results are presented in Section 5. Related work is discussed in Section 6 and Section 7 concludes the chapter.

http://dx.doi.org/10.5772/53772

2. Polling Service-Based MAC protocol

In this section, we briefly review PSMACs to provide the necessary background for the testbed. We refer interested readers to [9] for more technical details.

PSMAC is motivated by the insights from polling system theory [9]. Generally, a polling system consists of a shared resource (i.e., the wireless channel) and multiple stations (i.e., the wireless nodes). Polling systems may have either a centralized or a distributed structure. In the centralized case, a server maintains state information of the stations and polls the stations for channel access. In the distributed scenario, the stations contend for channel access using a distributed mechanism. In either case, one of the following three types of service policies can be used to serve the frames for a wining station: (i) *Exhaustive* policy, where a station is served until its buffer is emptied; (ii) *Gated* policy, where a station is served until all the frames that have backlogged in its buffer when the service begins are transmitted; (iii) *Limited-k* service, where a station is served for up to k frames or until the queue is empty, whichever comes first. It has been shown that both exhaustive service and gated service are more efficient than limited-k service, and they can guarantee bounded delay as long as the offered load is strictly less than 100% [14].

Based on the polling system theory, three polling service-based MAC protocols are introduced in [8, 9]. The main idea is to serve multiple frames after a successful contention resolution, thus amortizing the high control overhead over multiple DATA frames and making the protocols more efficient. The operation of PSMACs are shown in Fig. 3. In particular, PSMAC 1 senses a channel with CSMA/CA and uses RTS/CTS frames for contention resolution. All the frames to be transmitted are queued in a common transmission buffer. A winning node will use gated service to serve its backlogged frames. PSMAC 2 introduces multiple virtual queues, one for each neighbor. The gated service is used for one of the non-empty virtual queues when the station wins the channel. This allows other neighbors that are not involved in the transmission be scheduled to sleep for energy conservation. PSMAC 3 extends PSMAC 2 by serving all non-empty virtual queues when a station wins the channel, which may achieve even higher efficiency. Specifically, PSMAC 3 introduces a new control frame *announcement frame* (AF). AF is broadcasted after a sender wins the channel by RTS, which contains the lengths of all the non-empty virtual queues at the sender, as well as the order in which the virtual queues will be served. Thus, each neighbor will realize how many frames it will receive, as well as the starting and ending time for its reception. The sender then starts data transmission by clearing the virtual queues one by one by gated service in the order that announced by AF. The current receiving node is active for the reception, while all other neighbors can be scheduled to sleep and to wake up when its corresponding virtual queue is to be served.

All of PSMACs introduced are based on gated policy in polling theory. Exhaustive policy may achieve higher efficiency, however, it is not practically implementable. This is due to the fact that the new frames may arrive at the buffer after the transmission start. The source node can not determine the exact transmission time before sending RTS. Thus, extra coordination control protocols are needed for the scheduling.

In [8, 9], the PSMACs are evaluated with analysis and simulations. They are shown to achieve considerable throughput and delay improvements over p-Persistent CSMA, which is used as a proper benchmark for the performance evaluation due to its similarity to the IEEE 802.11 DCF [10]. In addition, PSMACs 2 and 3 can achieve significant energy savings

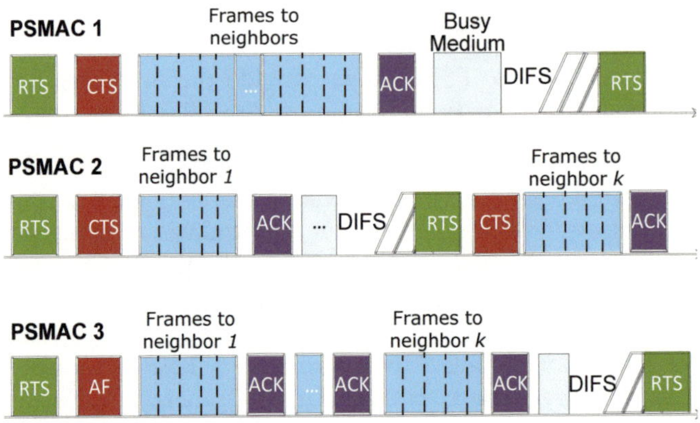

Figure 3. Timeline illustration of PSMACs operation.

by scheduling nodes to sleep, when they are not involved in the transmission of a packet train. The PSMACs are also shown to be more efficient for handling bursty traffic types and asymmetric traffic patterns, and the performance gains are achieved without sacrificing fairness performance [8, 9].

When $k = 1$, the limited-1 policy is a special case of limited-k, with only up to one frame served for a winning station. This policy is used in most existing MAC protocols, e.g., p-Persistent CSMA and IEEE 802.11 DCF. We focus on the PSMAC 2 protocol in this paper since it is most compatible to the DCF. We also implement a limited-1 based IEEE 802.11 DCF like protocol for performance comparison purpose. Both implementations are based on the GNU Radio/USRP platform [12, 13]. We call the PSMAC 2 and limited-1 MAC implementations GR-PSMAC and GR-Limited-1, respectively, in the rest parts of the paper (where GR stands for GNU Radio).

3. Testbed system overview

3.1. GR-PSMAC and GR-Limited-1

We implement GR-PSMAC and GR-Limited-1 by extending the IEEE 802.11 DCF, which is the de-facto protocol for WiFi networks. In particular, the implementations integrate CSMA/CA with binary exponential backoff, virtual carrier sense, RTS/CTS handshake, and ARQ for link error control to make full operational MAC protocols.

In GR-PSMAC, a station maintains multiple *virtual queues*, one for each of its neighbors. That is, DATA frames for different neighbors are enqueued into different virtual queues. When there is one or more non-empty virtual queues, the source station will selects a nonempty virtual queue in the *round-robin* fashion and start to sense the channel. After the channel is idle for DIFS interval, the *CW* start to decrease. If the channel remains idle when *CW* reaches 0, an RTS frame will be transmitted. If the channel is busy, *CW* will be frozen and the transmission will be deferred. When the destination station receives the RTS, it may return a CTS frame to confirm that it is ready for receiving data. The CTS frame contains the transmission duration, which may contains multiple frame durations and allows other stations set up their NAV for virtual carrier sensing. After receiving the CTS, the gated-service will be used for the selected virtual queue, i.e., the source station will transmit its backlogged DATA frames back-to-back to the destination following the gated service policy. All other stations will keep silent and wait for the NAV to expire (or, they may be scheduled to sleep for energy conservation). When the last frame is received, an ACK frame is issued by the target receiver to acknowledge all the successfully received frames, which will be removed from the virtual queue at the source station. If some frames are not correctly received after the transmission phase, the backoff procedure will be performed at the source station to defer the transmissions. The lost or corrupted frames will be retransmitted at a later time. This procedure is illustrated in Fig. 3.

The backoff procedure used in the implementation is illustrated in Fig. 2, which follows the IEEE 802.11 DCF specification. In this chapter, we set $CW_{min} = 8$ and $CW_{max} = 256$ as in [1]. After each successful transmission or when the number of RTS retries reaches a predefined maximum value, CW will be reset to CW_{min}.

GR-Limited-1 is implemented in the similar manner, except that when the source station wins the channel, only up to one DATA frame will be transmitted for a winning station (as shown in Fig. 1). This is consistent with the standard IEEE 802.11 DCF and its performance is comparable to IEEE 802.11 DCF and used for performance comparison with the proposed GR-PSMAC.

3.2. Software and hardware platforms

We develop the PSMAC testbed on the Software Defined Radio (SDR) platform consisting of GNU Radio and USRP [12, 13]. SDR is a modern approach to wireless communications [15], which allows dynamic reconfiguration of waveforms by software. GNU Radio [12] is an open-source software development toolkit under the GNU General Public License (GPL). It provides signal processing runtimes and processing blocks to implement SDR on RF hardware and commodity processors. GNU Radio applications are usually written in Python scripts, which allows the quick reconfiguration of the protocols, while the compiled C++ codes are used for the signal processing components of physical layer for minimal processing time. USRP [13] is a generic SDR hardware device that natively integrates with GNU Radio. We use USRP1 as the hardware platform for prototyping. The motherboard of USRP1 consists of four 64 MS/s ADCs and four 128 MS/s DACs. It has an FPGA for processing baseband and IF signals. The RFX2400 RF front-end daughterboard supports transmission and receiving from 2.3 GHz to 2.9 GHz in the ISM band. Integrated with USRP, GNU Radio provides a compelling software platform for prototyping wireless communications and networking protocols.

During the implementation, we observe that the main limitation of GNU Radio for MAC development is the high latency. Most MAC protocols rely on precise receiving and transmission timing. For example, IEEE 802.11 requires precise timing for the virtual carrier sensing mechanism. However, GNU Radio introduces a non-negligible latency due to the general-purpose processor and USB interface. In addition, the bus system to transfer the samples between a radio front-end and the processor also introduces extra latency. Finally, the Python script environment, kernel/user space switch and process scheduling of the operation system also make the latency hard to track. It is reported in [16] that the modulation, spreading, demodulation and despreading procedures could introduce an additional 22.5 ms delay, which is quite large comparing to the standard timing setting in IEEE 802.11 (generally in the μs scale).

Figure 4. The testbed wireless station setup.

The large latency also negatively affects performance measurement during testbed experiments, especially under high transmission rates. To tackle this problem, we use a relatively small link rate along with a large frame size to mitigate the impact of latency on transmissions. For example, using a 125 kbps link capacity with 1,500-byte frames, the frame transmission delay is about 96 ms, which is about 70% of the total transmission latency. With reduced link rates, we can conduct full functional tests for the MAC protocols and obtain precise *normalized* performance results with the given platform. It is worth noting that the Gigabit Ethernet interfance used in the later version of USRP, and the implementing the protocol functions in the FPGAs as in Rice University's Wireless Open-Access Radio Platform (WARP) platform [17], will help to allieviate the latency issue.

4. Testbed implementation description

We develop the MAC protocols on the GNU Radio/USRP platform [12, 13]. Each wireless station in the testbed consists of a USRP1 unit and a laptop (or desktop) computer, as illustrated in Fig. 4. We describe the implementation related issues in this section.

4.1. Network protocol architecture

Both GR-PSMAC and GR-Limited-1 are implemented as Layer 2 protocols from the point of view of network protocol architecture. Both protocols are written in Python scripts and are running in the user space of Linux. Since there is no explicit interface to directly access the MAC from the user space, we resort to the Linux TAP/TUN virtual network interface that provides the bridge between GNU Radio and Linux TCP/IP kernel. Specifically, we create a *virtual* Ethernet interface, termed gr_0, which can be configured with an IP address. Applications can then use the MAC protocols implemented in GNU Radio transparently as a standard network application programming interface (API). This approach is illustrated in Fig. 5.

To implement the MAC layer functions, we design the MAC header as given in Fig. 6, which is similar to that of IEEE 802.11. The header fields are defined as follows.

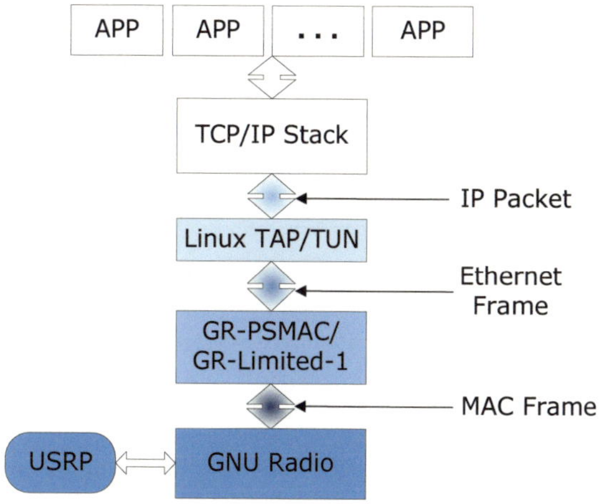

Figure 5. Protocol architecture of the GNU Radio testbed.

- Frame Control: four least significant bits define the frame type (RTS/CTS/DATA/ACK); other bits are reserved for future use.
- Destination Address: address of the destination node.
- Source Address: address of the source node.
- Next Hop Address: address of the next hop node; only valid for DATA frames and is used for the access point mode or multi-hop mode.
- Duration: multi-purpose field; in RTS/CTS/DATA: number of frames to be transmitted; in ACK: sequence number of the last received DATA frame.
- Sequence Number: sequence number of transmitted DATA frame; in ACK: sequence number of the first received DATA frame.
- Count: in RTS/CTS/DATA: number of transmitted frames; in ACK: number of correctly received DATA frames.
- Option: reserved for future use.

The PSMAC header contains eight fields and is 16-bytes long in total. Although some of the fields are compatible with the header definition of IEEE 802.11; the header format is different from the standard Ethernet header. For example, standard 48-bit MAC addresses are used for the Linux TAP/TUN frame, but two-byte addresses are used to identify the USRP hardware in PSMAC. Therefore, frames from the upper layer through the TAP/TUN driver will require a mapping from Ethernet header to PSMAC header, as illustrated in Fig. 5. Similarly, GR-PSMAC and GR-Limited-1 also map PSMAC headers back to the Ethernet header for received frames.

4.2. Transmission and receiving path

The GR-PSMAC is implemented as two execution data paths, namely, the *transmission path* and the *receiving path*. We adopt multithreading and each path is controlled by a thread. The design of the two paths is shown in Fig. 7 and outlined below.

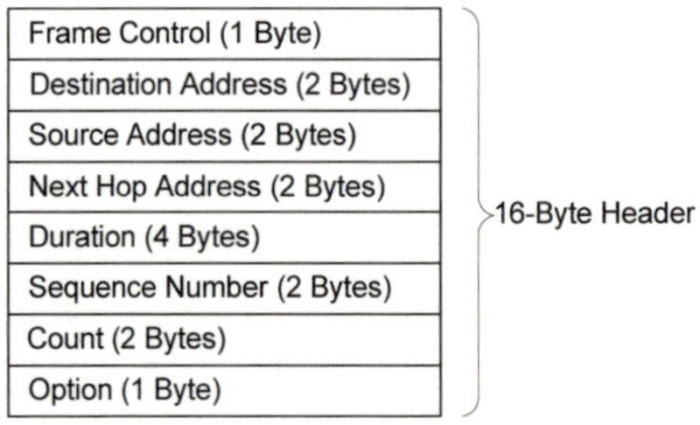

Figure 6. GR-PSMAC/GR-Limited-1 Header Format.

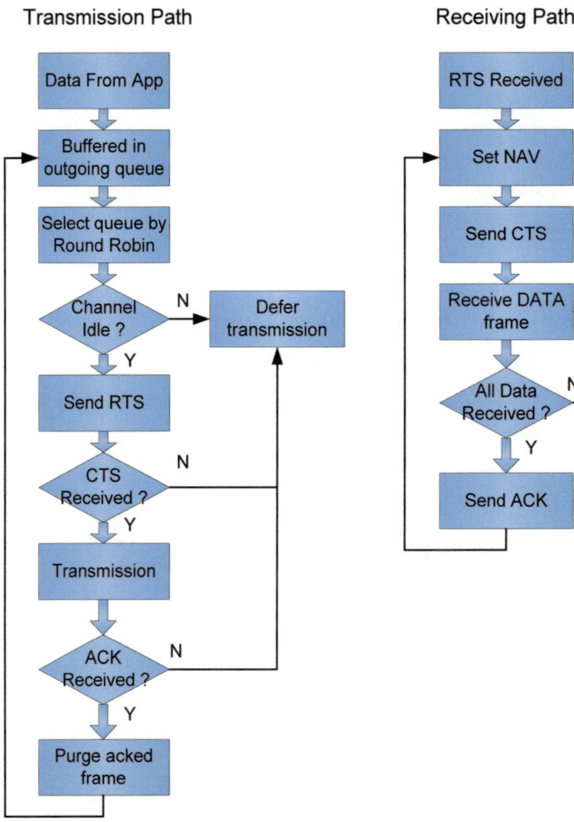

Figure 7. Illustration of the transmission and receiving operation.

4.2.1. Transmission Path

When GR-PSMAC receives a DATA frame from the upper protocol stack, it replaces the Ethernet header with the PSMAC header and buffers the frame in the outgoing queue. If the channel is sensed busy, the frame is held in the outgoing queue and the transmission is deferred. As discussed, GR-PSMAC maintains a virtual queue for each of its neighbors. The DATA frames are enqueued to the virtual queues according to their destination MAC addresses.

If the channel is sensed idle, the station selects a non-empty virtual queue in the round-robin manner, and issues an RTS frame to the neighbor corresponding to the chosen virtual queue.

http://dx.doi.org/10.5772/53772

The requested transmission time in the RTS is equal to the duration for transmitting all the backlogged frames in the selected virtual queue. If a CTS frame is not returned before timeout, GR-PSMAC will backoff the transmission and increase the RTS retry counter by one. Furthermore, if the RTS retry number exceeds a predefined limit, GR-PSMAC will reset *CW* and serve the next nonempty outgoing virtual queue for the fair operation among the virtual queues.

On the other hand, if a CTS frame is successfully received, GR-PSMAC will reset its *CW*, transmit the DATA frames in a row that had been backlogged in the selected virtual queue when the RTS was sent, and wait for ACK. If an ACK frame is received before timeout, GR-PSMAC will purge the acknowledged frames from the outgoing virtual queue. Otherwise, it will backoff the transmission and try to serve the next nonempty outgoing virtual queue.

4.2.2. Receiving Path

When a station receives an RTS destined for itself (i.e., carrying its MAC address as destination), it sets its NAV according to the *Duration* field value in the RTS. Then it returns a CTS frame with the duration equal to the original duration minus the CTS frame duration. Other neighbors that receive the CTS frame will set their NAV according to the *Duration* field and enter the sleep mode.

During the following transmission period, the destination station receives one or more back-to-back DATA frames. It maps the PSMAC headers back to Ethernet headers, and forwards the Ethernet frames to the upper layer. The sequence numbers of received DATA frames are recorded in a list. After all the frames are received or when there is a timeout, an ACK frame is issued with the successfully received sequence numbers back to the source station. The source station, once receiving the ACK, will remove all the successfully transmitted frames from its outgoing virtual queue.

Since both the transmission path (i.e., sending DATA frames) and receiving path (i.e., clearing acknowledged DATA frames) need to access the outgoing virtual queues, the multi-thread control need to be designed. A fast thread synchronization lock is introduced to protect the access conflict of the common resources of the transmission path and receiving path.

4.3. Acknowledge and retransmission mechanisms

We also implement the acknowledge and retransmission mechanisms for GR-PSMAC. IEEE 802.11 DCF uses limited-1 service that transmits only up to one DATA frame each time, such that a subsequent ACK acknowledges the successful DATA transmission. That is, a stop-and-wait ARQ mechanism is sufficient in this case. In GR-PSMAC, there may be multiple DATA frames transmitted in a row during the transmission period. Therefore a default ACK frame is not sufficient for acknowledging multiple DATA frames.

We implement two ARQ options for GR-PSMAC. The first one is Go-Back-N. The destination station records the received sequence numbers in the increasing order. When timeout happens or the last frame is received, the destination sends an ACK carrying the first received sequence number in the batch, as well as the last received sequence number right before the first missing frame (if any), by reusing the *Duration* field (see Fig. 6). All the frames received after the first missing frame will be discarded and retransmitted.

Although Go-Back-N ARQ is easy to implement, it is not efficient when the number of transmitted frames is large or when the frame loss rate is low. To improve efficiency and reduce the retransmission cost, we also implement the Selective Repeat Protocol (SRP). In SRP, the ACK issued by the destination node contains an explicit list of the sequence numbers of successful received frames; only the missing frames need to be retransmitted. SRP is generally more efficient than Go-Back-N protocol, because it can reduce the number of retransmissions, but with a slightly higher control overhead (i.e., longer ACK frames) and complexity.

4.4. Synchronization for distributed delay measurement

In a distributed network scenario, the CPU clocks may not be precisely synchronized. This may introduce frame delay measurement errors. To address the synchronization issue, we adopt the Precision Time protocol (PTP) daemon that implements the IEEE 1588 standard [18] to synchronize the testbed nodes. IEEE 1588 provides real-time clock synchronization for distributed systems with sub-microsecond precision. Such precision is sufficient for experiments and delay measurement in the PSMAC testbed.

We implement the delay measurement in the MAC layer as follows. The testbed nodes are connected with an Ethernet hub, and are then synchronized with the PTP daemon. When a DATA frame is enqueued at the source node, a time stamp will be stored at the source node. When a DATA frame is successfully received, the destination node will attach a time stamp in the ACK frame that records the time when the DATA frame was received, along with the list of sequence numbers. The MAC layer can directly monitor the outgoing queues and the event of frame receptions, which is free from the extra scheduling latency in the upper layers. The source station can compute the one-way delay as the difference between the received (i.e., in the ACK frame) and stored time stamps.

During the testbed experiments, we use the above mechanism frame delays for evaluation of the proposed schemes. For normal operation of the PSMAC implementation, however, such synchronization (and the Ethernet connections) is not required. Furthermore, with SRP ARQ, each ACK frame of GR-PSMAC carries more than one sequence numbers and timestamps of all the correctly received DATA frames, for the purpose of one-way delay measurement. In the normal operation mode, the ACK frame can be much shorter by carrying the sequence numbers of missing frames only, and by not carrying the timestamps. Therefore the control overhead could be further reduced and better throughput and delay performance could be achieved.

5. Experiments and results

5.1. Experiment setting

The GNU Radio PSMAC testbed consists of four USRP1 kits, each connected to a general purpose computer through a USB 2.0 port, as shown in Fig. 4. GR-PSMAC and GR-Limited-1 are implemented in GNU Radio 3.3 with Ubuntu Linux OS. As discussed, we also connect all the computers to an Ethernet hub and synchronize their clocks with IEEE 1588 (for accurate measurement of one-way frame delays).

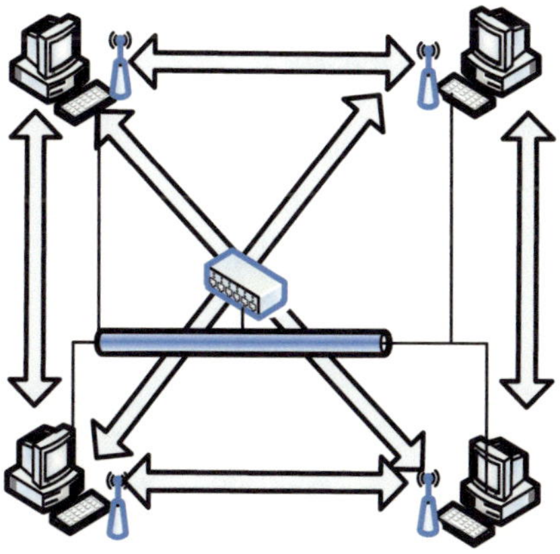

Figure 8. Network topology setups for the testbed experiments: ad hoc mode with a uniform traffic pattern.

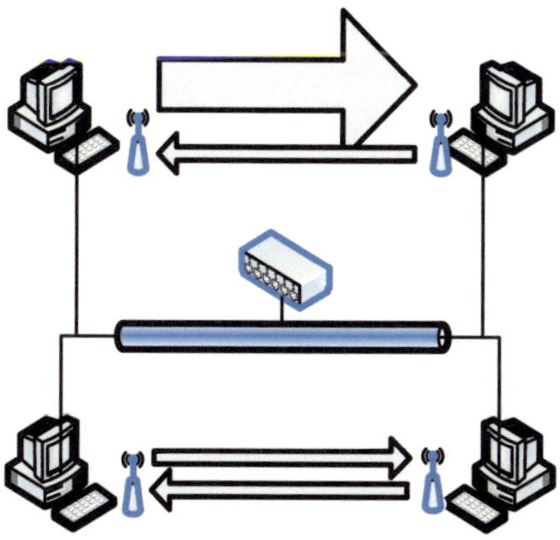

Figure 9. Network topology setups for the testbed experiments: ad hoc mode with a non-uniform traffic pattern.

We run the experiments using two network topologies as shown in Figs. 8–10: (i) a single-hop ad hoc network topology where each nodes communicates with every other node, as shown in Figs. 8 and 9, and (ii) an access point (AP) mode, where one station serves as the AP to relay traffic for the other three stations, as shown in Fig. 10. During the tests, the four testbed stations share the 125 kbps nominal link capacity. The more efficient SRP ARQ scheme and the RTS-CTS virtual carrier sensing mechanism are used. The default parameter settings are listed below.

- Samples per symbol: 2
- Carrier frequency: 2.401 GHz
- Modulation scheme: GMSK
- RTS Retry Limit: 5
- DATA frame size: 1,500 Bytes

We develop a UDP client-server application in C++ that can generate traffic to drive the experiments. UDP is chosen to avoid the complex rate variations caused by TCP congestion control, thus focusing on the MAC performance. The following three traffic models are used in the experiments:

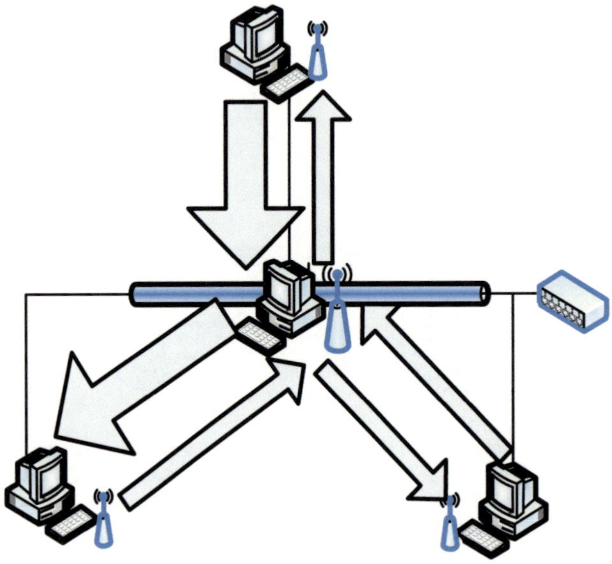

Figure 10. Network topology setups for the testbed experiments: AP mode with a non-uniform traffic pattern.

i) *i.i.d. Bernoulli* traffic: a frame is generated in each time slot with a predefined probability.

ii) *On-Off bursty* traffic: frames are generated according to an on-off Markovian model with geometrically distributed on and off periods. The average on period is five, while the average off period is tuned to achieve different offered loads, for the results reported in this section.

iii) *Long range dependent* (LRD) traffic: frames are generated according to an on-off traffic model with Pareto distributed on and off periods. It is shown that such source exhibit long range dependence [19]. The Hurst parameter is chosen to be $H = 0.7$ for the results reported in this chapter.

The first two traffic models belong to the class of *short range dependent* (SRD) models and are sufficient for modeling voice over IP traffic and the LRD model is a useful for modeling computer data traffic, which is shown to be self-similar [19]. The LRD model is much more bursty than first two traffic models, and the experiments with LRD traffic model take much longer time to converge to the steady state.

In the tests, we also consider different traffic patterns, by controlling the traffic rates at the source stations and the destination address of the generated DATA frames. With the *uniform* traffic pattern, the destination of each DATA frame is uniformly distributed among all the neighbors; with the non-uniform traffic pattern, one source-destination pair has much higher load than others.

For each offered load, we run the testbed experiment for ten times. Each experiment lasts for 300 s when the i.i.d. Bernoulli and On-Off bursty traffic models are used, and 3,000 s when the LRD traffic model is used. The offered load is increased from 0.1 to 1.0 in steps of 0.1 for the test scenarios. In the figures presenting experimental results, each point is the average of the ten tests, while the 95% confidence intervals are plotted as error bars.

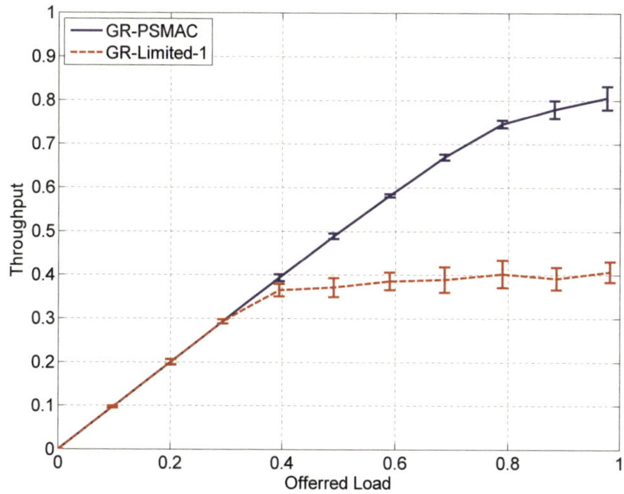

Figure 11. Normalized throughput under uniform i.i.d. Bernoulli traffic: ad hoc mode.

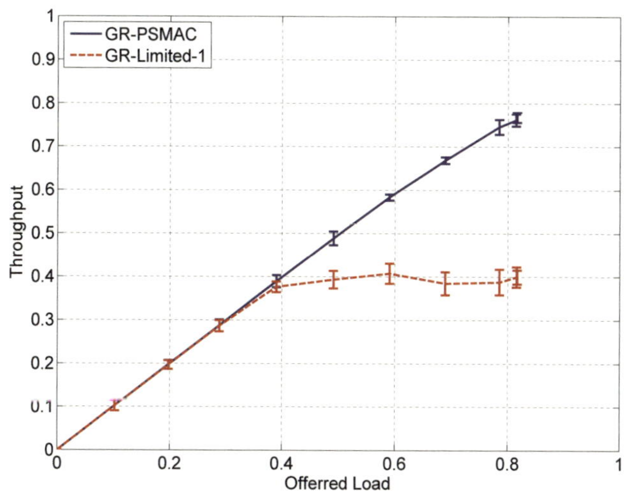

Figure 12. Normalized throughput under uniform On-Off busty traffic: ad hoc mode.

5.2. Experimental results

5.2.1. Performance under SRD Traffic: Ad Hoc Mode

5.2.1.1. Throughput and Delay

We first examine the network-wide throughput under the uniform Bernoulli and On-Off bursty traffic models and uniform traffic pattern. As shown in Fig. 8, each node uniformly sends UDP datagrams to all of its neighbors, and the offered loads for all the nodes are identical.

The network-wide normalized throughput performance are presented in Fig. 11 for the uniform Bernoulli traffic case and in Fig. 12 for the uniform On-Off traffic case. It can be seen that when the offered load is low, the achieved network-wide throughput is almost

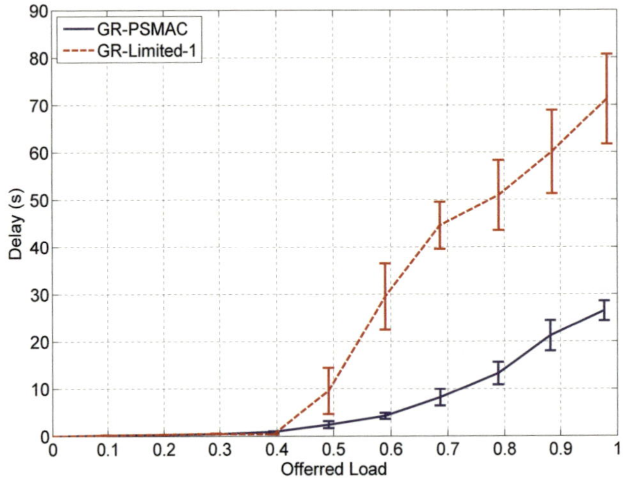

Figure 13. Average frame delay under uniform i.i.d. Bernoulli traffic: ad hoc mode.

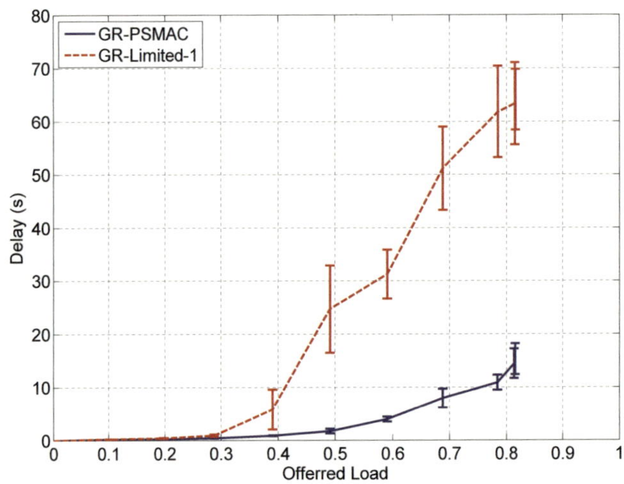

Figure 14. Average frame delay under uniform On-Off bursty traffic: ad hoc mode.

identical to the offered load. However, the normalized throughput saturates at about 40% when GR-Limited-1 is used in both Bernoulli and On-Off traffic cases, indicating congestion when the offered load exceeds 40%. On the other hand, the GR-PSMAC throughput keeps increasing even when the offered load is close to 100%. The maximum throughput of GR-PSMAC is about twice as high as that of GR-Limited-1.

We next evaluated the the frame delay under the same setup as in Fig. 8. The average delay for successfully received DATA frames are plotted in Figs. 13 and 14 for the uniform Bernoulli and On-Off traffic models, respectively. It can be seen that the GR-PSMAC delay is consistently much lower than the GR-Limited-1 delay for the entire range of offered loads. Under uniform Bernoulli traffic, the GR-PSMAC delay is only 37.16% of the GR-Limited-1 delay when the offered load is 98%. Under uniform On-Off bursty traffic, the GR-PSMAC delay is only 23.86% of the GR-Limited-1 delay when the offered load is 81.5%.

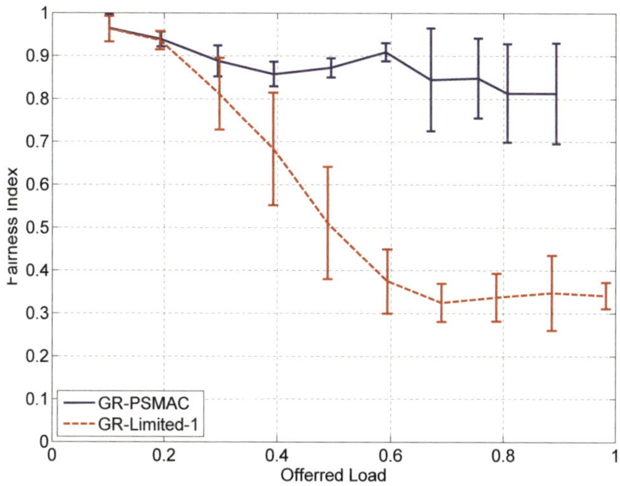

Figure 15. Fairness performance under non-uniform i.i.d. Bernoulli traffic: ad hoc mode.

5.2.1.2. Fairness

A common myth about gated or exhaustive polling service is that although the throughput/delay performance are superior, the fairness performance may not be good, since a heavily loaded node could use a larger fraction of the link capacity. To validate this common belief, we next examine the fairness performance with a non-uniform traffic pattern, as illustrated in Fig. 9. In this setting, the link from station 1 to station 2 takes 85% of the offered load, while the other 3 links share the remaining 15% offered load. Both the i.i.d. Bernoulli traffic and On-Off bursty traffic models are tested. We use the fairness index defined in [20]. For a system wit N stations, the fairness index is:

$$f(D_1, D_2, \cdots, D_N) = \frac{(D_1 + D_2 + \cdots + D_N)^2}{N(D_1^2 + D_2^2 + \cdots + D_N^2)},$$

where D_i is the average delay for the frames transmitted by station i, for $i = 1, 2, \cdots, N$. It can be verified that f is always between zero and one. In the fairest case, all the nodes have the same average delay, i.e., $D_1 = D_2 = \cdots = D_N$, and we have $f = 1$; in the worst case when one station's delay is dominant, i.e., $D_i >> D_j$, for all $j \neq i$, we have $f \approx 1/N$ (and $f = 0$ as $N \to \infty$).

The fairness indices achieved by GR-PSMAC and GR-Limited-1 are plotted in Figs. 15 and 16. It can be observed that all the fairness index curves drops as the offered load is increased, indicating the negative effect of congestion on fairness performance. In most cases, the GR-PSMAC fairness index is above 80% even under very high offered load, except for one point in the On-Off traffic case. On the other hand, the GR-Limited-1 fairness index curves drop to around 30% when the offered load exceeds 60% under both traffic patterns.

For further insights, we plot the per station average delay for GR-PASMAC and GR-Limited-1 under the non-uniform Bernoulli traffic in Fig. 17. We focus on the knee point when the offered load is 60%. It can be seen that with GR-PSMAC, every station has an average delay

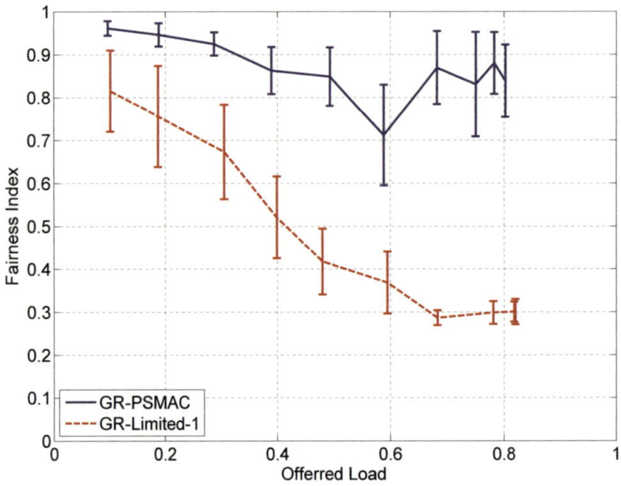

Figure 16. Fairness performance under non-uniform On-Off bursty traffic: ad hoc mode.

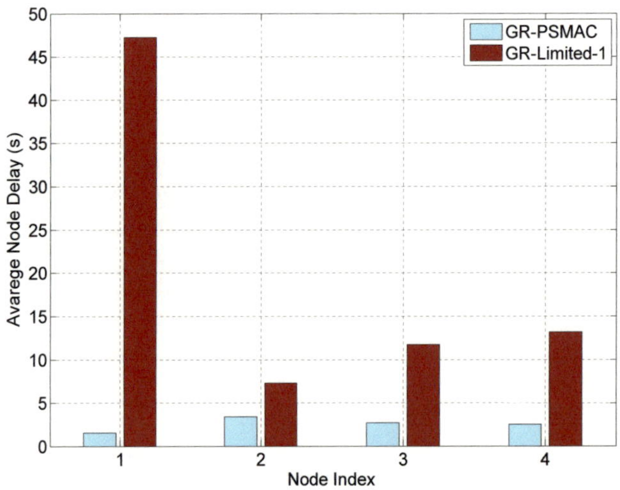

Figure 17. Average delay for each station under non-uniform i.i.d. Bernoulli traffic when the offered load is 60%: ad hoc mode.

smaller than 4 s. Although station 1 is transmitting at a rate 17 times as high as that of the other three stations, their average delays are close to each other, ranging between [1.60 s, 3.40 s]. Under GR-Limited-1, the heavily loaded station 1 has an average delay of 47.29 s, while the other three lightly loaded nodes have much lower average delays (all less than 14 s). GR-PASMAC achieves not only lower per station average delay than GR-Limited-1, but also more evenly distributed average delays among the stations than GR-Limited-1.

Therefore, the use of gated service in GR-PSMAC does not result in poor fairness. On the contrary, it achieves better fairness performance than limited-1 based schemes. This is largely due to the high efficiency and greatly reduced control overhead of PSMAC. All the virtual queues are efficiently served. The benefit introduced by gated service to a heavily loaded station does not significantly increase the delays of other lightly loaded nodes.

5.2.2. *Performance under LRD Traffic: Ad Hoc Mode*

In addition to i.i.d. Bernoulli and On-Off bursty traffic models, we also investigate the testbed performance under the LRD traffic model. It has been well known that computer data and VBR video traffic are *self-similar*, with Hurst parameters ranging from 0.5 to 1.0 [19]. For such traffic type, the class of SRD traffic models are inadequate to capture the complex autocorrelation structure. We adopt the On-Off traffic model with Pareto distributed on/off periods, which is an accurate model for LRD sources. By tuning the average duration of the off periods, the LRD process has a Hurst parameter of 0.7 for the experiments.

The simulation results with the LRD sources are presented in Figs. 18, 19, and 20. These results are obtained with the same topology and setting as the previous experiments, except that the traffic source is now the LRD source. In general, all the performance curves with the LRD sources have the same trend as those in the SRD case, and significant performance gains in throughput, delay and fairness are achieved by GR-PSMAC over GR-Limited-1.

Furthermore, GR-Limited-1 has worse performance under LRD traffic than that under SRD traffic. In Fig. 18, the throughput becomes saturated when the offered load exceeds 30%, which is earlier than the 40% offered load in the SRD case. The saturated throughput is 30%, which is also lower than the 40% saturated throughput in the SRD case. In Fig. 19, the SRD delay starts to diverge before the offered load reaches 30%, which is earlier than the SRD case in Fig. 14. When the offered load is 50%, for example, the GR-Limited-1 achieves an average delay of 76 s in the LRD case, a big increase from 25 s in the SRD case. In Fig. 20, the LRD fairness performance of GR-Limited-1 is similar to that in the SRD case, although the fairness index in the high offered load range is slightly lower than that in the SRD case. Such performance degradation clearly demonstrate the negative impact of LRD traffic on the network system performance.

On the other hand, we do not observe any performance degradation when the traffic sources are LRD. For example, the throughput curve in Fig. 18 is similar to that in Fig. 12, and the delay curve in Fig. 19 is also similar to that in Fig. 14, for the range of offered load examined. In Fig. 20, the GR-PSMAC fairness indices are all higher than 0.9, which is slightly better than that of the SRD case shown in Fig. 16.

Under LRD sources, it is more likely that the backlogged frames are concentrated in a small number of virtual queues. With gated service, such backlogged virtual queues can be quickly cleared out during one service period. Therefore GR-PSMAC is more effective in support LRD traffic, which has high rate variations and are generally very difficult to management and control.

5.2.3. *Performance in the AP Mode*

Finally, we examine the performance of GR-PASMAC with an infrastructure-based wireless network topology. As shown in Fig. 10, one station is configured to operate as the AP and the remaining three stations WLAN nodes that communicate with each other through the AP. With PSMAC, each WLAN node maintains a single outgoing queue, since the field of next hop address in all outgoing frame header is fixed to the address of the AP. The AP buffers the incoming packets in three different virtual queues, one for each of the WLAN nodes. We configure a nonuniform traffic pattern and use the On-Off burst traffic model for this star topology. Specifically, the traffic flow from station 1 to station 2 takes 30% of the offered load,

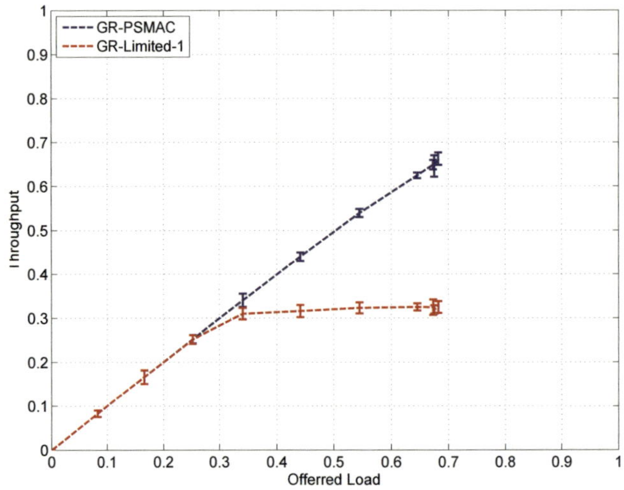

Figure 18. Normalized throughput under uniform LRD traffic: ad hoc mode.

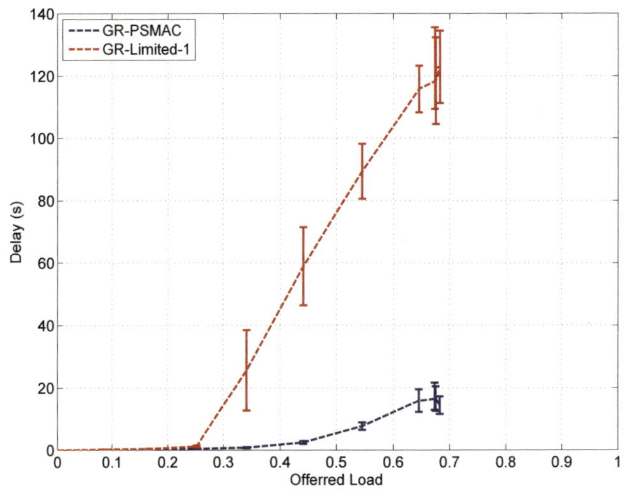

Figure 19. Average frame delay under uniform LRD traffic: ad hoc mode.

while the traffic flow from station 2 to station 3 and the traffic flow from station 3 to station 1 takes 10% of the offered load, respectively. Since the AP relays traffic, each frame will be transmitted twice. This scenario can also represents a multihop wireless topology, in which all the none-AP nodes are two hops from each other, and the AP becomes a hotspot of the multihop network.

In Fig. 21, we plot the normalized throughput for the AP topology. We observed that GR-PSMAC still achieves considerably higher throughput under different offered loads, while the throughput of GR-Limited-1 becomes saturated when the offered load exceeds 40%. The average frame delays are plotted in Fig. 22. It can be observed that the GR-PSMAC delays are consistent with the previous experiments, while the GR-Limited-1 delay shoots up when the offered load exceeds 30%. In Fig. 23, the GR-PSMAC fairness indices are constantly above 0.8, while the GR-Limited-1 fairness index curve drops when the offered load exceeds

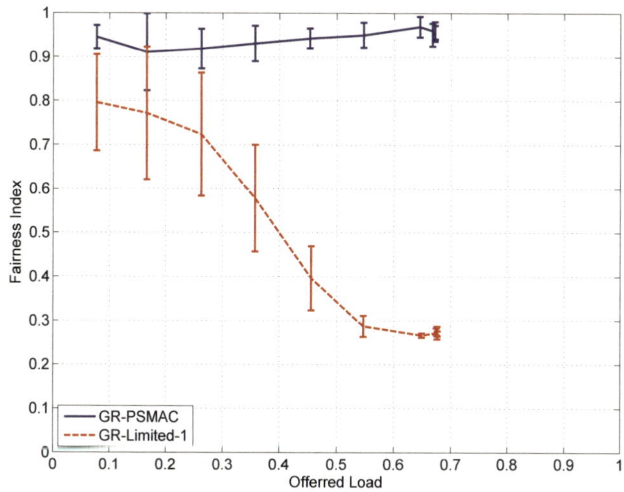

Figure 20. Fairness performance under non-uniform LRD traffic: ad hoc mode.

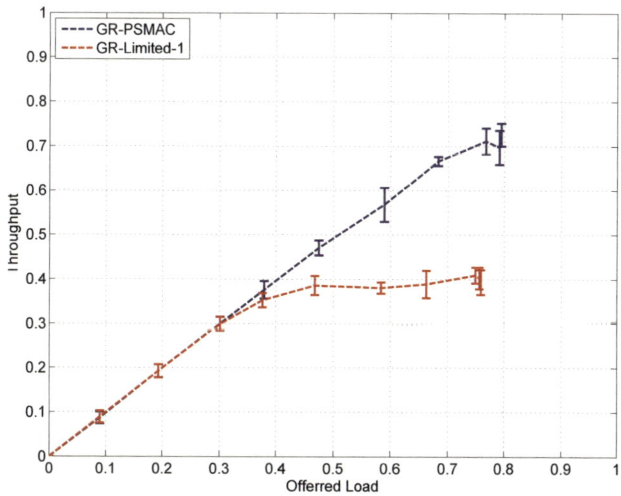

Figure 21. Normalized throughput with On-Off bursty non-uniform traffic: AP mode.

20%, to about 0.4 for the high offered load region. All these AP results are consistent with those for the ad hoc topology.

We further plot the average backlog length of the virtual queues at the AP in Fig. 24. The offered load is 60%. The ith virtual queue stores frames to be transmitted by the AP to the ith WLAN node. The average virtual queue backlogs for GR-PSMAC are 10.4, 25.5, and 2.3, while the GR-Limited-1 virtual queue backlogs are 78.5, 462.8, and 86.9. Clearly, the gate service incorporated in PSMAC is much more effective in clearing the backlogs at the AP node. The high efficiency of GR-PSMAC brings about significant benefits to alleviate congestion at the wireless hotspot.

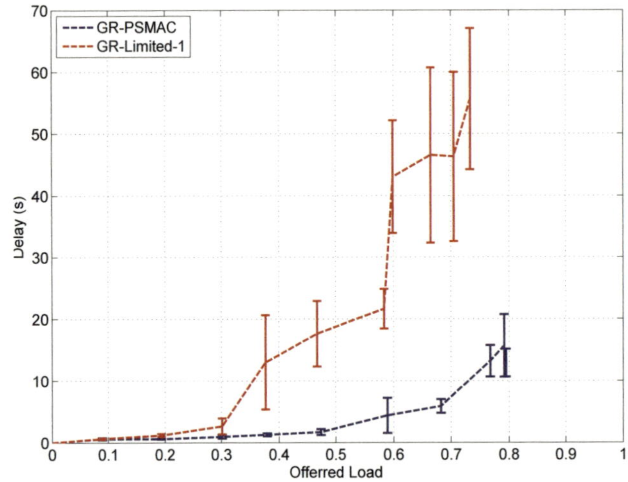

Figure 22. Average frame delay with On-Off bursty non-uniform traffic: AP mode.

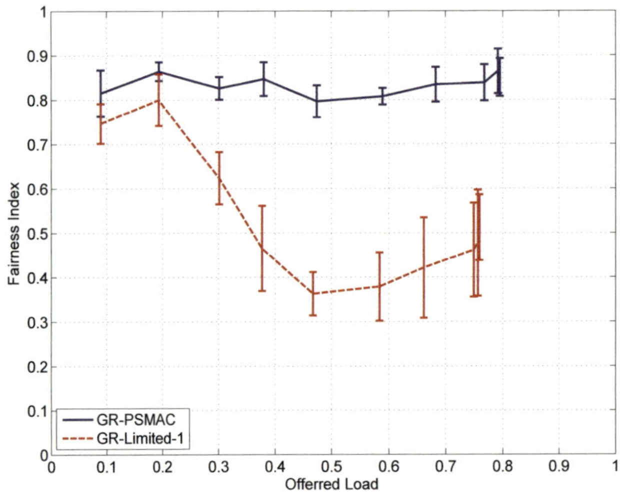

Figure 23. Fairness performance with On-Off bursty non-uniform traffic: AP mode.

6. Related work

This work is closely related to the research effort on improving the efficiency of MAC protocols. There are many techniques that exploit new capabilities of the wireless system to achieve this goal, such as adopting downlink MIMO (DL MIMO) [21], exploiting new spectrum allocation [22–24], exploiting spectrum opportunities in underutilized licensed band [25, 26], and exploit location information to schedule concurrent transmissions [27–29]. Exploiting efficient scheduling of the transmissions is a useful approach that is complimentary to the above techniques.

In particular, PSMAC that incorporates gated or exhaustive services was first introduced in [8, 9]. Limited-*k* service was used in the CM MAC [30], where *k* is equal to the *concatenated threshold*. In IEEE 802.11e HCF (hybrid coordination function) controlled channel access

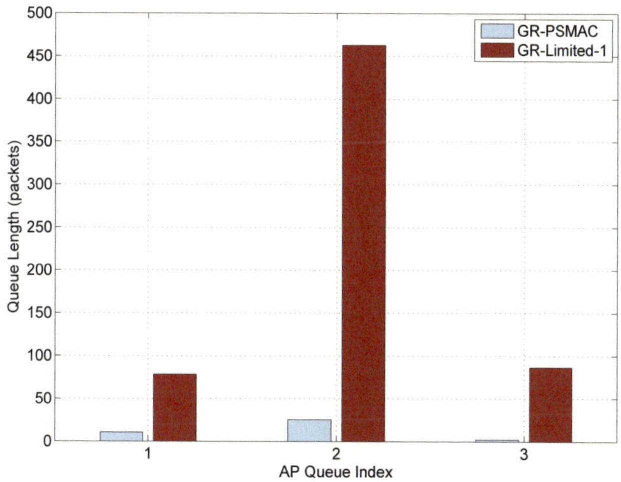

Figure 24. Virtual Queue lengths at the AP with On-Off bursty non uniform traffic when the offered load is 60%: AP mode.

(HCCA), the HC (Hybrid Coordinator, i.e., the AP) can assign Transmit Opportunities (TXOP) to a station, to allow the station send multiple frames in a row. This is a centralized approach originally designed to support real-time applications with regular traffic patterns, but the specific service discipline or algorithm for determining how many TXOPs to assign to a node, are not specified. In addition, a centralized controller is required to poll the secondary nodes, which is different from the random access and fully distributed approach taken in this work.

GNU Radio/USRP is a popular platform for prototyping wireless systems. In [31], the authors discuss the general implementation issues of the prototyping of wireless systems with USRP and GNU Radio. A main branch of prototyping works focus on the PHY, due to the configurable signal processing ability offered by GNU Radio [32–40]. In [32], an implementation of a MIMO PHY is reported. In [33], the authors developed a new wireless carrier sensing approach termed *LinkSense* to obtain fine-grain indications of channel activity. *LinkSense* utilizes a few OFDM subcarriers for conveying the link signature in each symbol, enabling sensing of active links at any time instant. The feasibility of *LinkSense* is then demonstrated on the GNU Radio/USRP platform with an OFDM implementation. In [34], the authors presented a software-defined IEEE 802.11b receiver and channel impulse response (CIR) measurement system. A USRP and GNU Radio testbed is designed to validate the CIR measurement system. The match filters are implemented in FPGA, while the Python code collects data from the USB, demodulates the packet, and records results.

In [35], the authors verify a multihop, multirate adaptation mechanism with a small scale USRP/GNU Radio-based testbed, while in [36], an implementation of Angle-of-arrival-assisted Relative Interferometric (ARI) RADAR transceiver is proposed based on GNU Radio and USRP. In [37], USRP, GNU Radio and OSSIE [41] are integrated to prove the concept of Government Reference Architecture (GRA), which is a standard for establishing a modular open system architecture for a family of Above 2GHz (A2G) Tactical Military Satellite Communications (MILSATCOM) terminals operating over several radio frequency bands. In [38], a cooperative communication testbed for both single-relay cooperation and multi-relay cooperation was reported based on GNU Radio and USRP2.

The significant performance enhancement for link reliability and end-to-end throughput of cooperative transmissions were observed. In addition, USRP and GNU Radio are also used to facilitate the prototype of RF front-end hardware. In [39], an RF front-end with 50 MHz – 2.5 GHz frequency range is designed and tested.

Due to the flexibility and full accessibility to the PHY and MAC, GNU Radio/USRP platform has been used to prototype MAC protocols that exploit PHY features [42–46]. The Hydra project [42] is a flexible wireless network testbed developed at UT Austin. The project exploits the Click modular router [47], GNU Radio, and C++ codes to prototype a cross-layer design of a rate adaptive MAC protocol. CoopMAC [43] is a programmable cooperative communication testbed developed at Polytechnic Institute of NYU. The testbed implements cooperative protocols in both PHY and MAC layers on the GNU Radio/USRP platform. The testbed experiment results verified significant benefits of cooperation in wireless networks. In [44], a load-adaptive MAC protocol is designed that switches between CDMA and TDMA based on traffic loads. Its performance is evaluated with a MIMO MANET testbed implemented with USRP based SDR nodes. In [45], the authors studied the performance of the IEEE 802.11 MAC under channel-oblivious and channel-aware jamming by theoretical analysis and extensive simulations via a GNU Radio/USRP testbed. An 802.11b ad hoc network with UDP traffic flows is established, where the sender, the receiver and the jammer are all implemented with USRP and GNU Radio. In [46], Dhar et al. presented a simple framework for joint design of MAC and PHY layers with the GNU Radio and Click platform. In [47], a software framework is presented, in which GNU Radio functions are encapsulated as a single Click element to provide PHY layer functionality. Due to the primarily goal of GNU Radio for supporting signal processing, the functions of MAC protocol are not fully supported. One of the main concern in MAC prototyping is precise timing in carrier sensing. The latency of GNU Radio/USRP brings about a significant challenge for high speed data rates. This issue is is analyzed in [16], among others. The transmit and receive latencies were evaluated and the impact on network performance was characterized under an IEEE 802.15.4 implementation.

As the interest in Cognitive Radio (CR) networks increases, GNU Radio/USRP has become popular in developing CR systems [48–51]. In [48], an adaptive interference avoidance Transform-domain Communication System (TDCS) based cognitive radio was demonstrated. In [49], GNU Radio/USRP is used to set up a testbed for service discovery and device identification in CR networks, which may achieve better spectral efficiency and also enhance wireless security. In [50], a cognitive receiver is designed, which includes a universal classifier, synchronizer, and demodulator. The performance is verified with the GNU Radios/USRP platform together with MATLAB-enabled Anritsu MS2781A Signature Signal Analyzer. In [51], the authors implemented an adaptive spectrum sensing scheme that exploits primary network traffic information with GNU Radio/USRP. USRP is shown to be amenable to implementing spectrum sensing algorithms.

In this book chapter, we focus on the fast prototyping of the PSMACs to evaluate the performance under the realistic wireless channels and networks. It could be possible to extend this work to more efficient industrial product by the recently developed frameworks. For example, besides GNU Radio and USPR, FPGA-based software radio platform, such as Airblue [52] was also designed to support high performance wireless protocols and cross-layer experiments. A very recent study promotes the wireless MAC processor concept [53], which provides engines for reconfigurable MAC protocol implementation.

The processor defines the programming interface through actions, events and conditions to support full-custom MAC protocol programming. The effectiveness of the wireless MAC processor are evaluated by AirForce54G chipset and proved that the processor can be implemented over an ultra-cheap commodity WLAN card.

7. Conclusion and future directions

In this book chapter, we presented the design and implementation of PSMAC, a gated service based MAC protocol, as well as a limited-1 based IEEE 802.11 DCF like MAC for comparison purpose. The testbed was developed on the GNU Radio/USRP platform. We discussed related design issues on the prototyping process. In addition, we also presented extensive experimental results under various traffic models and traffic patterns. The experimental study validated the analysis and simulation studies presented in our prior work, and demonstrated the advantages of PSMAC under a realistic wireless network setting. In future, it would be interesting to extend this work by integrating the PSMACs into IEEE 802.11 framework, such as *mac802.11* framework [54] and *Madwifi* [55] in Linux kernel, for a realistic addition to the industrial wireless MAC standard.

Acknowledgements

This work is supported in part by the US National Science Foundation (NSF) under Grants ECCS-0802113 and DUE-1044021, and through the NSF Wireless Internet Center for Advanced Technology at Auburn University. Any opinions, findings, and conclusions or recommendations expressed in this material are those of the author(s) and do not necessarily reflect the views of the foundation.

Author details

Yingsong Huang[1], Philip A. Walsh[2],
Shiwen Mao[1] and Yihan Li[1]

1 Department of Electrical and Computer Engineering, Auburn University, Auburn, AL, USA
2 QUALCOMM Inc., San Diego, CA, USA

References

[1] IEEE. Wireless LAN media access control (MAC) and physical layer (PHY) specifications, 2007.

[2] IEEE. Part 11, Wireless LAN Medium Access Control (MAC) and Physical Layer (PHY) Specifications: Amendment 5: Enhancements for Higher Throughput, Oct. 2009. IEEE Std 802.11n.

[3] IEEE. Part 11, Wireless LAN Medium Access Control (MAC) and Physical Layer (PHY) Specifications: Medium Access Control (MAC) Enhancements for Quality of Service (QoS), July 2003. ANSI/IEEE Std 802.11e, Draft 5.0.

[4] Bluetooth Special Interest Group, "Specification of the Bluetooth system," Nov. 2003, version 1.2. [online]. Available: http://www.bluetooth.com.

[5] M. Ozdemir, D. Gu, A. B. McDonald, and J. Zhang. Enhancing MAC performance with a reverse direction protocol for high-capacity wireless LANs. In *Proc. VTC'06 Fall*, pages 1–5, Montreal, Canada, Sept. 2006.

[6] Y. Xiao and J. Rosdahl. Throughput and delay limits of IEEE 802.11. *IEEE Commun. Lett.*, 6(8):335–357, Aug. 2002.

[7] J. Li, C. Blake, D. D. Coute, H. Lee, and R. Morris. Capacity of ad hoc wireless networks. In *Proc. ACM MOBICOM'01*, pages 61–69, Rome, Italy, July 2001.

[8] Y. Li, S. Mao, and S. Panwar. PSMAC: Polling service-based medium access control for wireless networks. In *Proc. WiOpt'07*, pages 1–10, Limassol, Cyprus, Apr. 2007.

[9] Y. Li, S. Mao, S. Panwar, and S. Midkiff. On the performance of distributed polling service-based medium access control. *IEEE Trans. Wireless Commun.*, 7(11):4635–4645, Nov. 2008.

[10] R. Bruno, M. Conti, and E. Gregori. Optimization of efficiency and energy consumption in p-persistent CSMA-based wireless LANs. *IEEE Trans. Mobile Computing*, 1(1):10–31, Jan./Mar. 2002.

[11] Y. Huang, P. Walsh, Y. Li, and S. Mao. A GNU radio testbed for distributed polling service-based medium access control. In *Proc. IEEE MILCOM 2011*, pages 519–524, Baltimore, MD, Nov. 2011.

[12] GNU Radio, [online] Available: http://www.gnu.org/software/gnuradio.

[13] Ettus Research LLC, "USRP family brochure." [online] Available: http://www.ettus.com/downloads/.

[14] H. Takagi and L. Kleinrock. A tutorial on the analysis of polling systems. Computer Science Department, UCLA, Tech. Rep. No. 850005, Feb. 1985.

[15] Jeffrey H. Reed. *Software Radio: A Modern Approach to Radio Engineering*. Prentice Hall, May 2002.

[16] T. Schmid, O. Sekkat, and M. Srivastava. An experimental study of network performance impact of increased latency in software defined radios. In *Proc. WinTECH'07*, pages 59–66, Montreal, Canada, Sept. 2007.

[17] WARP: Wireless Open-Access Research Platform, Rice University, [online] Available: http://warp.rice.edu/index.php.

[18] IEEE. IEEE 1588TM-2002 standard for a precision clock synchronization protocol for networked measurement and control systems. [online] Available: http://www.nist.gov/el/isd/ieee/ieee1588.cfm.

[19] K. Park. The Internet as a complex system. In K. Park and W. Willinger, editors, *SFI Studies in the Sciences of Complexity*, chapter 1. Oxford University Press, Oxford, UK, 2005.

[20] R. Jain, D.M. Chiu, and W. Hawe. A quantitative measure of fairness and discrimination for resource allocation in shared systems, 1984. DEC Research Report TR-301.

[21] M. X. Gong, E. Perahia, R. J. Stacey, Want R, and Mao S. A CSMA/CA MAC protocol for multi-user MIMO wireless LANs. In *Proc. IEEE GLOBECOM 2010*, pages 1–6, Miami, FL, Dec. 2010.

[22] M. X. Gong, R. J. Stacey, D. Akhmetov, and S. Mao. Performance analysis of a directional CSMA/CA protocol for mmWave wireless PANs. In *Proc. IEEE WCNC 2010*, pages 1–6, Sydney, Australia, Apr. 2010.

[23] M. X. Gong, D. Akhmetov, R. Want, and S. Mao. Directional CSMA/CA protocol with spatial reuse for mmWave wireless networks. In *Proc. IEEE GLOBECOM 2010*, pages 1–5, Miami, FL, Dec. 2010.

[24] I. K. Son, S. Mao, M. X. Gong, and Y. Li. On frame-based scheduling for directional mmWave WPANs. In *Proc. IEEE INFOCOM 2012*, pages 2149–2157, Orlando, FL, Mar. 2012.

[25] D. Hu and S. Mao. Design and analysis of a sensing error-aware MAC protocol for cognitive radio networks. In *Proc. IEEE GLOBECOM 2009*, pages 5514–5519, Honolulu, HI, Nov./Dec. 2009.

[26] D. Hu and S. Mao. A sensing error aware MAC protocol for cognitive radio networks. *ICST Transactions on Mobile Communications and Applications*, 1(1), 2012. to appear.

[27] S.-M. Hur, S. Mao, Y. T. Hou, K. Nam, and J. H. Reed. On exploiting location information for concurrent transmission in multi-hop wireless networks. *IEEE Transactions on Vehicular Technology*, 58(1):314–323, Jan. 2009.

[28] I. K. Son, S. Mao, and S.-M. Hur. Medium access control for opportunistic concurrent transmissions under shadowing channels. *MDPI Sensors Journal*, 9(6):4824–4844, June 2009.

[29] S.-M. Hur, S. Mao, K. Nam, and J. H. Reed. On concurrent transmissions in multi-hop wireless networks with shadowing channels. In *Proc. IEEE ICC 2008*, pages 2662–2666, Beijing, P.R. China, May 2008.

[30] Y. Xiao. IEEE 802.11 performance enhancement via concatenation and piggyback mechanisms. *IEEE Trans. Wireless Commun.*, 4(5):2182–2192, Sept. 2005.

[31] D.C. Tucker and C.A. Tagliarini. Prototyping with GNU radio and the USRP - where to begin. In *Proc. IEEE SOUTHEASTCON '09*, pages 50–54, Mar. 2009.

[32] X. Li, W. Hu, H. Yousefi'zadeh, and A. Qureshi. A case study of a MIMO SDR implementation. In *Proc. IEEE MILCOM'08*, pages 1–7, San Diego, CA, Nov. 2008.

[33] N. Santhapuri, R.R. Choudhury, and S. Nelakuditi. Link sense: beyond wireless carrier sensing. *IEEE Trans. Commun. Letters*, 15(4):470–472, Apr. 2011.

[34] D. Maas, M. Firooz, J. Zhang, N. Patwari, and S. Kasera. Channel sounding for the masses: low complexity GNU 802.11b channel impulse response estimation. *IEEE Trans. Wireless Commun.*, PP(99):1–8, Nov. 2011.

[35] C. Yu, T. Shen, K.G. Shin, J. Lee, and Y. Suh. Multihop transmission opportunity in wireless multihop networks. In *Proc. IEEE INFOCOM'10*, pages 1–9, San Diego, CA, Mar. 2010.

[36] J. Friedman, A. Davitian, D. Torres, D. Cabric, and M. Srivastava. Angle-of-arrival-assisted relative interferometric localization using software defined radios. In *Proc. IEEE MILCOM'09*, pages 1–8, Boston, MA, Oct. 2009.

[37] T. Leising, C. Dietrich, M. Gavitt, R. Kim, H. Satake, S. Hoque, and T. Rittenbach. An excursion to define the boundaries of the Government Reference Architecture. In *Proc. IEEE MILCOM'08*, pages 1–6, San Diego, CA, Nov. 2008.

[38] J. Zhang, J. Jia, Q. Zhang, and E. Lo. Implementation and evaluation of cooperative communication schemes in software-defined radio testbed. In *Proc. IEEE INFOCOM'10*, pages 1307–1315, San Diego, CA, Mar. 2010.

[39] M. Bruno, M. Murdy, P. Perreault, A.M. Wyglinski, and J.A. McNeill. Widely tunable RF transceiver front end for software-defined radio. In *Proc. IEEE MILCOM'09*, pages 1–6, Boston, MA, Oct. 2009.

[40] P. Fuxjager, D. Valerio A. Costantini, P. Castiglione, G. Zacheo, T. Zemen, and F. Ricciato. IEEE 802.11p transmission using GNURadio. In *Proc. WSR'10*, pages 83–86, Karlsruhe, Germany, Mar. 2010.

[41] J. Snyder, B. McNair, S. Edwards, and C. Dietrich. OSSIE: an open source software defined radio platform for education and research. In *Proc. FECS'11*, pages 18–22, Las Vegas, NV, July 2011.

[42] K. Mandke, C. Soon-Hyeok, K. Gibeom, R. Grant, R. Daniels, K. Wonsoo, R. Heath, and S. Nettles. Early results on Hydra: A flexible MAC/PHY multihop testbed. In *Proc. IEEE VTC'07*, pages 1896–1900, Dublin, Ireland, Apr. 2007.

[43] T. Korakis, M. Knox, E. Erkip, and S. Panwar. Cooperative network implementation using open-source platforms. *IEEE Commun. Mag.*, 47(2):134–141, Feb. 2009.

[44] W. Hu, H. Yousefizadeh, and X. Li. Load adaptive MAC: A hybrid MAC protocol for MIMO SDR MANETs. *IEEE Trans. Wireless Commun.*, 10(11):3924–3933, Nov. 2011.

[45] E. Bayraktaroglu, C. King, X. Liu, G. Noubir, R. Rajaraman, and B. Thapa. On the performance of IEEE 802.11 under jamming. In *Proc. IEEE INFOCOM'08*, pages 1265–1273, Phoenix, AZ, Apr. 2008.

[46] R. Dhar, G. George, A. Malani, and P. Steenkiste. Supporting integrated MAC and PHY software development for the USRP SDR. In *Proc. 2006 1st IEEE Workshop on Networking Technologies for Software Defined Radio Networks*, pages 68–77, Sept. 2006.

[47] E. Kohler, R. Morris, B. Chen, J. Jannotti, and M.F. Kaashoek. The Click modular router. *ACM Trans. Comput. Syst.*, 18(3):263–297, Aug. 2000.

[48] R. Zhou, Q. Han, R. Cooper, V. Chakravarthy, and Z. Wu. A software defined radio based adaptive interference avoidance TDCS cognitive radio. In *Proc. IEEE ICC'10*, pages 1–5, Cape Town, South Africa, May 2010.

[49] R. Miller, W. Xu, P. Kamat, and W. Trappe. Service discovery and device identification in cognitive radio networks. In *Proc. IEEE SECON'07*, pages 40–47, San Diego, CA, Jun. 2007.

[50] Q. Chen, Y. Wang, and C.W. Bostian. Universal classifier synchronizer demodulator. In *Proc. IEEE IPCCC'08*, pages 366–371, Austin, TX, Dec. 2008.

[51] X. Shi and R. de Francisco. Adaptive spectrum sensing for cognitive radios: an experimental approach. In *Proc. IEEE WCNC'11*, pages 1408–1413, Quintana-Roo, Mexico, Mar. 2011.

[52] M. C. Ng, K. E. Fleming, M. Vutukuru, S. Gross, A. Arvind, and H. Balakrishnan. Airblue: A system for cross-layer wireless protocol development. In *Proc. ANCS'10*, pages 1–11, La Jolla, CA, Oct. 2010.

[53] I. Tinnirello, G. Bianchi, P. Gallo, D. Garlisi, F. Giuliano, and F. Gringoli. Wireless MAC processors: programming MAC protocols on commodity hardare. In *Proc. IEEE INFOCOM'12*, pages 1269–1277, Orlando, FL, Mar. 2012.

[54] Linux Wireless, [online]. Available: http://linuxwireless.org/.

[55] Madwifi Project, [online]. Available: http://madwifi-project.org/.

Section 2

Routing Protocols for Wireless Ad-Hoc Networks

Chapter 3

Graph-Based Routing, Broadcasting and Organizing Algorithms for Ad-Hoc Networks

Li Liu, Xianyue Li, Jiong Jin, Zigang Huang,
Ming Liu and Marimuthu Palaniswami

Additional information is available at the end of the chapter

http://dx.doi.org/10.5772/54146

1. Introduction

The development of networks of low-cost, low-power, multi-functional devices has received increasing attention over the last ten years. These devices are small in size and able to process data, communicate with each other, typically over a radio channel, and even sense. Each device particpates in a self-configuring infrastureless network connected by wireless, called ad-hoc network. Since most of the individual node in ad-hoc networks is inherently resource constrained: limited processing speed, storage capacity, and communication range and energy, it is impossible to achieve application requirements by individual device or unattached devices. A number of devices within a network have to combine as an aggregate collaborating to achieve application requirements. However, such massive devices cooperation must be achieved by the necessary organizational structures without requiring human intervention.

An ad-hoc network is able to arrange itself to achieve the application requirements according to the present situations. Hence, wireless communication has to be the primary means to enable information exchange among these devices. In a wired network like the Internet, each router connects to a specific set of other routers, forming a routing graph. In ad-hoc networks, each device has a radio that provides a set of communication links to nearby devices. Multi-hop communication is expected to overcome some of the signal propagation effects experienced in long distance wireless communication.

In a wide array of disciplines, an ad-hoc network can be intuitively casted into the format of a graph which is a set of vertex and a set of edges that might connect pairs of the nodes. The ad-hoc network consists of devices (vertex or nodes) and the communication links (edges) between them. Graphs are seemingly ubiquitous in ad-hoc network field. The problems of designing multi-hop routing, broadcasting and organization algorithms for ad-hoc networks have received great considerable attention [22][23][24][25][26][27]. All are tightly coupled to the problem of the distinguished graphs. In this chapter, we discuss the routing, broadcasting and organization algorithms and protocols that can be formulated by three types of graph.

- Connected Dominating Set: Connected dominating set is useful in the computation of routing, broadcasting and organization for mobile ad-hoc networks. In mobile ad-hoc networks, each device is free to move independently, and therefore change its communications links to other devices frequently. A small connected dominating set is used as a backbone for communications. Nodes that are not in the connected dominating set communicate by passing messages through neighbors that are in this set.
- Disjoint Sets: Disjoint sets are used in the implementation of energy efficent routing and organization, e.g., scheduling nodesą̃ő status between running and sleeping, as well as in the aspect of fault tolerant routing. In mobile ad-hoc networks, several disjoint sets every pair of which have no nodes in common can gurantee multiple choices of message passing paths and nodes organization.
- Minimum Spanning Subgraph and Steiner Minimum Tree: Mininum spanning subgraph and Steiner minimum tree represent a spanning subgraph or a tree with the lowest total costs. The generation of subgraphs and Steiner trees has applications in mobile ad-hoc routing and organization design. Several varieties of the minimum spanning subgraph problem and steiner tree problem are proposed for the sake of describing the issues on the fault tolerant, topology control and constrained routing protocol design in mobile ad-hoc networks.

Most of these problem are either *NP-hard*. Several approximate and near-approximate algorithms are proposed to solve these issues based on the combinatorial optimization and graph theory. In real mobile ad-hoc networks, there are some restricted conditions to be achieved in various applications, which will make the problems more difficult to solve. In this chapter, we attempt to give a preliminary review of the design and implementation of the heursitic or approxiamte algorithms on routing, broadcasting and organization by using the combinatorial optimization and graph theory. Note that we only focused on the three problems, all of which were our previous research works. The interested reader is also referred to some excellent works on other topics of combinatorial optimization and graph theory[20][21].

The organization of this chapter is as follows. In Section 2, we give some basic definitions of graph theory that appear in ad-hoc network formulation. We also give the notations used throughout this chapter. In Section 3, we provide the main ideas and approaches of formulating the ad-hoc network issues into several versions of connected dominating set problems. We present our previous research works of proposed algorithms and results related to graph theory. Then in Section 4, we consider the ad-hoc network issues which can be formulated to find disjoint sets. We also present a method of converting disjoint sets issues to network flow and combinatorial optimization problems. In Section 5, we present the minimum spanning subgraph and minimum steiner tree problem applied in fault-tolerant algorithm design. Finally, Section 6 concludes this chapter.

http://dx.doi.org/10.5772/54146

2. Basic definitions and notations

An ad-hoc network topology could be represented by a graph G that is an ordered triple $(V(G), E(G); C(G))$ or $G(V, E; C)$ consisting of a nonempty set V of vertices $v_1, v_2, ..., v_n$, and a set E of edges, and C is the set of weights on the nodes or the edges. Generally, an edge denotes that the two nodes belong to it can communicate. Therefore,$E = \{(v_i, v_j) : dist(v_i, v_j) \leq r_i\}$, where $dist$ is the Euclidean distance function and r_i represents the transmission range of node i. The edge (v_i, v_j) denotes v_i is able to communicate with v_j. An unweighted graph G is also presented as $(V(G), E(G))$ or $G(V, E)$.

The graph G could be a directed graph if the network is heterogeneous that nodes have various transmission ranges, or be an undirected graph if any two nodes can communicate with each other where an edge (v_i, v_j) indicates that there must be an edge (v_i, v_j) existing in E. The weight on a node or an edge could denote the metrics of the network. In a power aware application, the weight might be the remaining power of a node. It might be a vector containing transmission speed and power consumption on an edge for the application that aims to find an energy-efficient delay-constraint routing path. In some applications, the weights on all nodes or edges are the same, e.g, in a fault tolerant network that needs achieve no requirement except finding routing paths between two nodes.

A graph H is a subgraph of G (written $H \subseteq G$) if $V(H) \subseteq V(G)$ and $E(H) \subseteq E(G)$. An induced subgraph of $G, G[V']$, contains a vertex set V', where V' is a nonempty subset of $V(G)$, and an edge set E', where $E' \subseteq E(G)$ that have both ends in V'. The induced subgraph $G[V(G) \backslash V']$ is denoted by $G - V'$. If $V' = \{v\}$, we write $G - v$. Similarly, an edge-induced subgraph of $G, G[E']$, contains a vertex set V' and an edge set E', where E' is a nonempty subset of $E(G)$ and ends of edges in E' belong to V'. The spanning subgraph of G with edge set $E(G) \backslash E'$ is written simply as $G - E'$. The graph obtained from G by adding a set of edges E' is denoted by $G + E'$. If $E' = \{e\}$, we write $G - e$ and $G + e$ instead of $G - \{e\}$ and $G + \{e\}$.

Let G_1 and G_2 be subgraphs of G. G_1 and G_2 are disjoint if they have no vertex in common, and edge-disjoint if they have no edge in common. The union $G_1 \cup G_2$ of G_1 and $G - 2$ is the subgraph with the vertex set $V(G_1) \cup V(G_2)$ and the edge set $E(G_1) \cup E(G_2)$; if G_1 and G_2 are disjoint, their union can be also denoted by $G_1 + G_2$.

The degree of a vertex v in G, $d(v)$, is the number of edges of G incident with v. $\delta(G)$ and $\Delta(G)$ represent the minimum and the maximum degrees of vertices of G respectively.

A path in G is a finit non-null sequence $P = v_0e_1v_1e_2v_2...e_kv_k$, whose terms are alternately vertices and edges, such that, for $l \leq i \leq k$, the ends of e_i are v_{i-1} and v_i. In addition, the vertices $v_0, v_1, ..., v_k$ are distinct, P is called a path. Usually, denote the section $v_iv_{i+1}v_j$ of the path $P = v_0v_1...v_k$ by $P[v_i, v_j]$. Two vertices u and v of G are connected if there is a $P[u, v]$ in G. u and v are directly connected or adjacent if $(u, v) \in E(G)$. Connection is an equivalence relation on the vertex set $V(G)$. Therefore, there is a partition of $V(G)$ into nonempty subsets $V_1, V_2, ..., V_k$ such that every pair of the vertices u and v is connected if and only if both u and v belong to the same set V_i. The subgraphs $G[V_1], G[V_2], ..., G[V_k]$ are called the components of G. If G has exactly one component, G is connected; otherwise, G is disconnected.

A flow network is a directed graph $G(V, E; C, f, s, t)$, where every edge $(u, v) \in E$ has a non-negative capacity $c(u, v)$, f is a flow function $f : V \times V \rightarrow \Re^+$, s is a source and t is a sink. A flow network must contain the properties for all nodes u and v: (1) The flow along an edge (u, v) cannot exceed its capacity $f(u, v) \leq c(u, v)$; (2) The flow to a node is zero, except for the source s, which produces flow, and the sink, which consumes flow, $\sum_{w \in V} f(u, w) = 0$, where $u \neq s$ or t.

3. Connected Dominating Set

A Connected Dominating Set (*CDS*) of an ad-hoc network is a subset of nodes in the network, where the nodes in *CDS* are responsible for maintaining routing information, and other nodes have to rely on these nodes in *CDS* for transmission. Exploring *CDS* problem is frequently used to model the problem of computing a minimum number on the set. *CDS* plays a very important role in routing, broadcasting and connectivity management in wireless ad-hoc and sensor networks where there is no pre-defined physical backbone infrastructure to support routing and topology control that makes routing-related tasks or hierarchical organizations are very complicated.

The *CDS* problem can be formulated as follows: a graph $G(V, E)$, a Dominating Set(*DS*) ia a subset $U \subseteq V$ such that for every vertex $v \in V$, either $v \in V$, or there exists an edge $(u, v) \in E$ and $u \in U$. If the induced subgraph $G[U]$ is connected, then U is called a *CDS*. The *CDS* problem is to find a *CDS* with minimum size. In this chapter, we will give three classes of this problem. Minimum Connected Dominating Set, which is the complementary problem of all *CDS* related problems, finds a set with minimum number of nodes to construct a virtual backbone or elect cluster heads in practice. Minimum Weighted Connected Dominating Set, where the graph is weighted on node that represents energy, cost, or neighbor size in real applications, finds the minimum sum of the weighted nodes to achieve better power consumption requirement. Fault Tolerant Connected Dominating Set finds a minimum set of nodes such that it remains a connected dominating set after any part of nodes leave, to guarantee the stability and robust of a backbone or a cluster-based network upon the node failure that occurs frequently in ad-hoc networks.

3.1. Minimum Connected Dominating Set

It is well-known that to find a Minimum Connected Dominating Set (*MCDS*) in a general graph is *NP-complete*. In wireless ad-hoc and sensor networks, if all nodes are homogeneous, Unit Disk Graph (*UDG*) is used to represent their geometrical structures. A *UDG* can be formally defined as follows: Given an undirected graph $G(V, E)$, each vertex v has a transmission range with radius 1. Two vertices u and v adjacent if their Euclidean distance is less than or equal to 1. Clark et al. [1] show that computing *MCDS* is also *NP-hard* in *UDG*, and a lot of approximation algorithms for *MCDS* can be found in the literature.

To find an approximated *MCDS*, the most popular method is as follows. Firstly, find a maximal independent set(*MIS*) in given graphs. Given a graph $G(V, E)$, an Independent Set(*IS*) is a subset $I \subset V$ such that for any two vertex $u, v \in I$, they are not adjacent, say, $(u, v) \notin E$. An *IS* is called a Maximal Independent Set if any other arbitrary vertex is added to this set, the new set will not be an *IS* any more. Compared with *CDS*, *MIS* is much easier to be constructed. Usually, we use $mis(G)$ to denote the size of the constructed *MIS*. The second step is to make this *MIS* connected. We donote the number of the added vertices in this step by $conn(G)$. Let $mcds(G)$ be the size of minimum *CDS*. Then, the approximation ratio for such algorithm is

$$\frac{mis(G)+conn(G)}{mcds(G)} = \frac{mis(G)}{mcds(G)} + \frac{conn(G)}{mcds(G)}$$

3.1.1. CDS in UDG

For the connecting part, the best-known algorithm is a Steiner tree based algorithm with $conn(G) \leq 3mcds(G)$ till now[7]. On the other hand, for selecting *MIS* part, there exist many results. Let M be the set of *MCDS*. Based on the geometry structure on *UDG*, if we increase$V \backslash M$ from 1 to 0.5, then we can construct a new graph G'. It is easy to see that all the disks in V are located insides the area formed by M. Then we can get a conclusion that the sum of maximum area for MIS should be less than the area of MCDS, which is a rough bound for $\frac{mis(G)}{mcds(G)}$.The following theorem gives this bound.

Theorem1.[2] The rough bound for $mis(G)$ and $mcds(G)$ is

$$mis(G) \leq 3.748mcds(G) + 5.252$$

Next, because the above result is rough, we analyzed the relationship between $mis(G)$ and $mcds(G)$ more specifically. Firstly, we used Voronoi Division to divide the whole area into some small Voronoi cell. The following Fig.1. gives an example. We also analyzed the area for each kind of polygons under densest situations. Then we can have a better bound for $\frac{mis(G)}{mcds(G)}$.

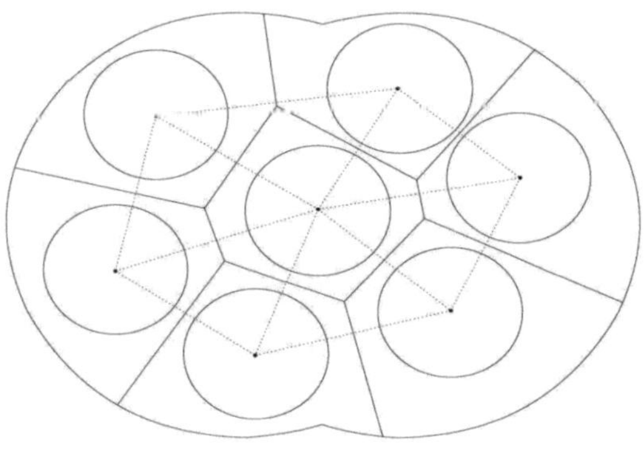

Figure 1. An example of Voronoi cell.

Theorem2.[2]$mis(G) \leq 3.453mcds(G) + 4.839$

Finally, we apply graph theory to consider the problem. By modifying the Voronoi division 3-regularization and combining classical Euler's Formula, we obtain the following result.

Theorem3.[2]$mis(G) \leq 3.399mcds(G) + 0.0790k + 4.874$, *where k is the number of the holes in the whole area.*

By these theoritical results, if we use two-step method to find a *CDS*, the size of the *CDS* is at most 6.4 times optimal solution. Now, we go back to Theorem 3, it has a parameter k. Then, the future work of this problem is to decide the parameter k. There exists litter rusults about it and the following theorem gives a basic solution.

Theorem4.[2]*For any unit disk graph G, let MCDS be a minimum connected dominating set. To form a hole, there need at least 6 connect vertices in MCDS.*

3.1.2. CDS in UBG

UDG is widely used to abstract the homogeneous wireless networks. However, sometimes, this assumption is far from the reality. In this case, to abstract homogeneous wireless networks in three-dimensional space, Unit Ball Graphs (*UBGs*) model is used. Since *UDGs* are special instances of *UBGs* in which the altitude of every node is the same, every *NP-hard* problem in *UDGs* is also *NP-hard* in *UBGs*. Naturally, *MCDS* in a *UBGs* is still *NP-hard*.

Like the *CDS* problem on *UDGs*, we use two-step method to find an approximated *MCDS*. The first step is to construct a *MIS* and to give a bound for $\frac{mis(G)}{mcds(G)}$ on *UBGs*. Recall the famous Gregory-Newton Problem concerning about kissing number, the kissing number $k(S_3) = 12$, that is, there all at most 12 independent unit balls that can simultaneously touch the surface of one unit ball. Based this result, there is a trivial bound $mis(G) \leq 11mcds(G) + 1$. In order to get a better result, we consider the problem: how many independent unit balls can simultaneously touch the surface of two adjacent unit balls. Through some accurate computation, we obtain the following lemma.

Lemma 1.*The number of independent nodes in the union of two adjacent unit balls is at most 22.*

Since the result in Lemma 1 is better than kissing number, we can improve the above ratio.

Theorem 5.[5]$mis(G) \leq 10.917mcds(G) + 1.083$

Let M be a *MIS* in a *UBG* such that for any partition (M_1, M_2) of M, $dist(M_1, M_2) = 2$. Next, we present two differernt greedy algorithms to connect M and give the approximation ratios.

Algorithm 1.[10]*Greedy Algorithm for CDS on UBGs.*
1: $H = G[M]$, *that is,* H *is the subgraph of* G *induced by* M;
2: **WHILE** H *is disconnected* **DO**
3: *Choose the vertex* v *which connected the maximum component of* H;
4: $M = M \cup v$ *and* $H = G[M]$;
5:**END WHILE**
6:**RETURN** M;

Theorem 6.[10]*The Algorithm 1 outputs a CDS in the unit ball graph G. And the size is up-bounded by* $(13 + ln10)opt + 1$, *where opt is the size of MCDS.*

Before introducing the Algorithm 2, some useful notations are presented. For any vertex x, let $N(x)$ be the set of vertices adjacent to x. For any vertex set U, let $N(U) = (\cup_{x \in U} N(x)) \backslash U$ and $M_{v,U}$ be the set of vertices which is adjacent to v and belong to $M \backslash U$.

Algorithm 2.[5]*CDS Computation Algorithm on UBGs*
1: $U = r$, *and* $M' = M - r$
2: **WHILE** $M' \neq \emptyset$ **DO**

3: *Choose the vertex* v *such that* $M_{v,U} = \max\{|M_{x,U}|\,|x \in N(U)\}$;
4: $U = U \cup v \cup M_{v,U}$ *and* $M' = M' \backslash M_{v,U}$;
5:**END WHILE**
6:**RETURN** U;

Theorem 7.[5]*The Algorithm 2 outputs a CDS in the unit ball graph G. And* $|U| \leq 14.937opt + 1.083$.

Using above two algorithms, we can get an approximated *CDS* in any *UBG G* with linear performance.

3.1.3. Multi-hop CDS in UDG

If we further consider the architecture of wireless networks, we can separate the network into many clusters and the selected *CDS* are cluster heads and gateways. Each node will send message to its local cluster head, and information is exchanged among those cluster heads through more steady and responsible channels, which makes the whole network more reliable.

For a *CDS*,each cluster is really small, including nodes only one hop away from the corresponding cluster head. Therefore, some researchers enlarged the size of clusters, such that the super cluster head can be at most d-hop away from the nodes within its dominating range. The set of such super cluster head is called *d-CDS*. Given a graph G, an *d-CDS* is a subset $U \subset V$such that for any vertex $v \in V \backslash U$, there exists a path form $v$to some vertex of U with length at most d. Furthermore, the subgraph induced by U is connected. d-hop *CDS* problem is also NP-complete for general graphs and *UDGs*. The Fig.2. gives an example for *2-CDS (TCDS)*.

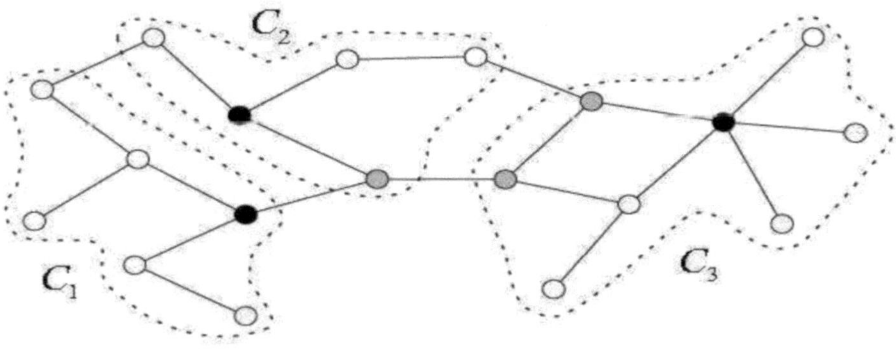

Figure 2. The black and gray vertices form a *TCDS* of G with three subgraphs C_1,C_2, C_3.

As the *CDS* problem, approximation algorithms for *d-CDS* problem usually divide into two steps. The first step, find a *d-DS* (usually *d-MIS*) in the graph. The second step, connect the *d-DS* into *d-CDS*. A *TCDS* algorithm can be generalized to *d-CDS* algorithm directly. Table 1 shows the local variables presented in Algorithm 3.

Then, we presented a distributed approximation algorithm for*TCDS* problem.

Name	Explanation
level	Number of hops from root to this node. The rootąŕs level=0.
id	The ordering number of this node.
rank	The list of neighbors within 2 hops ordered by (level, id).
color	Totally for colors, black, brown, grey and white.
blacklist	The path from one black node to another black node which actives it.

Table 1. Local Variables in Algorithm 3.

Algorithm 3.[3]*4-Coloring Algorithm for TCDS.*
Phase 1
1. Choose an arbitrary root r, set r's level as 0 and build a spanning tree T for r, such that each node will get its level information.
2. Each node exchange information to get a rank list, which records its neighbors within 2-hops.
Phase 2
1. Root r colors itself as black, and broadcasts a black message with its id.
2. When a node receives a black message, it broadcasts a brown message with path as (id_1, id_2), where id_1 is the sender's id and id_2 is the receiver's id. Besides, if this node is white or grey, mark itself as brown.
3. When a node receives a brown message with (id_1, id_2), it broadcasts a grey message with a path as (id_1, id_2, id_3), where id_3 is the receiver's id. If it is a white node, mark itself as grey.
4. When a node receives a grey message, if the node has been colored before, it will transfer this GREY message as T-grey message with sender's id. If the node is white, it gives labels to its colored neighbors in rank list. After update, if every of its lower-level neighbors have been colored already, record path as blacklist, mark itself as black, and broadcast a black message with its id.
5. When a white node receives a T-grey message with an id, it gives label to corresponding neighbor with same id in rank list as colored neighbor.
6. If a leaf node has been colored, it transmits a colored message to its parent. If a node has been colored and it receives colored message from all its children, it sends a colored message to its parent. Phase 2 will terminate until the root receives colored messages from all its children.
Phase 3
1. All black nodes join themselves into TCDS list, and send Join message with blacklist.
2. When receive a Join message, if a node's id is in blacklist, mark itself into TCDS set, and transmit this message until it reaches the black node with the first id in this list. The algorithm will terminate when all the nodes in blacklist has been inserted into TCDS set.

Using the same method in *CDS* problem on *UDG* to analysis the algorithm, we can obtain the approximation ratio of Algorithm 3.

Theorem 8.[3]*The Algorithm 3 outputs a TCDS in the UDG G with time complexity $O(n)$ and message complexity $O(nlgn)$. Furthermore, the result from Phase 2 has size at most* $5.807opt + 17.152$*, and Algorithm 3 has approximation ratio* 17.421.

Algorithm 3 can be easily modified into a distributed algorithm for *d*-hop *CDS* problem on *UDG* with approximation ratio $0.225r^3 + 1.337r^2 + 0.585r$, where $r = d + 0.5$. Hence, for any *UDG* and any fixed parameter *d*, we can get a *d*-hop *CDS* with constant approximation ratio.

http://dx.doi.org/10.5772/54146

3.2. Node-Weighted Connected Dominating Set

The Node-Weighted Connected Dominating Set (*NWCDS*) problem is a generalization of the *CDS* problem. Given a graph $G(V, E)$ with node weight function $f : V \to R^+$, the *NWCDS* problem is to find a *CDS* of G such that its total weight is minimum. For convenience, the weight function f such that $f(V) \geq 1$ is normalized for any vertex v in G. If the weights on all vertices are the same, the *NWCDS* problems are equal to the *CDS* problem. Hence, the *NWCDS* problems are also *NP-complete* on general graphs and *UDGs*.

To deal with this problem, we firstly considered the Min-Weight Chromatic Disk Cover (*MWCDC*) problem and used dynamic program to obtain a polynomial algorithm for *MWCDC*. Then, comparing the two problems and studying the relationship between them, we had the Lemma 2.

Lemma 2.[11]*If there exists an ρ-approximation algorithm for the MWCDC and for any fixed ε, there is a polynomial $(4\rho + \varepsilon)$-approximation algorithm for the NWDS.*

Based on Lemma 2 and the exact algorithm for *MWCDC*, we can get a $(4 + \varepsilon)$-approximation algorithm for *NWDS*. Then, using the $(1 + \varepsilon)$ -approximation algorithm for the Node-Weighted Steiner tree (*WST*) problem to connect the *NWDS*, we can obtain an approximation algorithm for *NWCDS* on *UDG*.

Theorem 9.[11]*There is a $(5 + \varepsilon)$-approximation algorithm for the MWCDS by using a node-weighted Steiner tree to interconnect all nodes of the MWDS.*

3.3. Fault-Tolerant Connected Dominating Set

In wireless ad-hoc and sensor networks, nodes are mobile and thus the topology of such networks can be changed frequently. As a result, a virtual backbone (*VB*) induced by a*CDS* can be broken easily and thus it should be re-computed repeatedly. Hence, to construct a fault-tolerance *VB* is important. Here, a k-connected m-dominating set is introduced as a generalized abstraction of a fault tolerant *VB*. Given a graph $G(V, E)$, a subset $U \subset V$ is a m-dominating set (m-DS) if for any vertex $v \in V \backslash U, v$ has at least m neighbors in U . Furthermore, if U is k-connected, we call U is a k-connected m-dominating set ((k, m)-*CDS*).

To obtain a (k, m)-*CDS*, the main idea is as follows. The first step is to get a basic *CDS*, that is, $(1, 1)$-*CDS*. Next, add $m - 1$ *MISs* in the rest graph to make the $(1, 1)$-*CDS* into $(1, m)$-*CDS*. Finally, by adding some new vertices to increase the connectivity of the *CDS*, the (k, m)-*CDS* is obtained. The first and second steps are easy to complete, but the final step is very hard for $k \geq 3$. When $k = 2$, there are some approximation algorithms, and the best one is given by Shang et. al. [8] with approximation ratio $5 + \frac{25}{m}$ for $2 \leq m \leq 5$ and 11 for $m > 5$.

In the following, we introduce a $(3, m)$-*CDS* approximation algorithm. The key idea about this algorithm is to become the "bad-points" in the $(2, m)$-*CDS* to good. Given a 2-connected graph G, a vertex v is called a "good-point" if $G - v$ is also 2-connected; otherwise, v is called a "bad-point". Based on above definition, we can get Lemma 3.

Lemma 3.[4]*A 2-connected graph without any bad-point is 3-connected.*

Algorithm 4.[4]*Algorithm for $(3, 3)$-CDS on UDGs.*
1: Computer a $C_{2,3}$, and Set $Y = C_{2,3}$;
*2: **WHILE** Y has bad-points **DO***

3: Choose an arbitrary bad-point v and Set $B = Y$;
4: Construct the leaf-block tree T of $B - \{v\}$ with blocks $\{B_1, B_2, ..., B_s\}$ and cut-vertices $\{c_1, c_2, ..., c_t\}$;
*5: **WHILE** there exist a non cut-vertex w in some block B_i and a cut-vertex c_j such that (w, c_j) is a saparetor of Y **DO***
6: Set $B = B_i$ and $v = c_j$;
7: Construct the leaf-block tree T of $B - \{v\}$ with blocks $\{B_1, B_2, ..., B_s\}$ and cut-vertices $\{c_1, c_2, ..., c_t\}$;
*8: **END WHILE***
9: Find a path H to make a bad-point $c_j \in \{c_1, c_2, ..., c_t\}$ to be a good-point such that all other new vertices of H are good;
10: Set $Y = Y \cup H$;
*11: **END WHILE***
*12: **RETURN** Y;*

Theorem 10.[4]*The Algorithm 4 output a (3,3)-CDS with approximation ratio* 520/3.

Easy to see, if we start the Algorithm 4 with a $C_{2,m}$ with $m \geq 3$, the algorithm will return a $(3, m)$-*CDS*. And we can get the following result.

Theorem 11.[4]*There exists a constant ratio approximation algorithm for $(3, m)$-CDS problem in UDG for any m.*

As above, given a *UDG* and an integer m, we can get a $(3, m)$-*CDS* as a fault-tolerance *VB* with constant ratio.

4. Disjoint Sets

Disjoint Sets (*DS*) of an ad-hoc network is a collection of disjoint sets of nodes that each set is capable of achieving application requirements. For instance in a wireless sensor network where coverage is an important demand, the Connect Disjoint Set divides the nodes into a number of disjoint sets, such that every set completely covers all the target points. Connected Disjoint Set problem is frequently used to formulate the problem into finding a maximum number of sets.

DS problem plays an important role in ad-hoc networks, especially in wireless sensor networks. Disjoint Set Covers (*DSC*) problem is one of the classical problems that aims to determine a maximum number of disjoint covers, where every cover is a set of sensors which together monitor all the target points. It can be formulated as a graph, and solved by combinatorial optimization method-mixed integer programming. Besides, we studied further to find maximum disjoint sets for maintaining not only coverage but also connectivity. This class of Disjoint Set problem could be used for node scheduling methods to conserve energy, topology control methods to tolerant failure, and routing protocol design.

The disjoint set covers (*DSC*) problem was addressed by M. Cardei and D.Z. Du [12] in order to solve the problem of energy efficiency for surveillance of a set of targets. Let $T = \{t_1, t_2, ..., t_m\}$ be a set of m targets. Each node covers a subset of targets. A collection of nodes $S = \{s_1, s_2, ..., s_n\}$ are defined as that each set $S_i = \{t_{i_1}, t_{i_2}, ..., t_{i_l}\}$ if node S_i covers targets $t_{i_1}, t_{i_2}, ..., t_{i_l}$. *DSC* problem aims to find a maximum number of disjoint sets of nodes, where every set is able to cover all the targets. They presented a heuristic algorithm based on

http://dx.doi.org/10.5772/54146

the flow network since the *DSC* problem was proved *NP-complete*. First, a bipartite directed graph $G(V,E)$ is constructed with the vertex set $V = S \cup T$ and $s_i t_{i_j} \in E$ with a capacity 1 if and only if $t_{i_j} \in s_i$. Then, draw k copies of G, namely $G_1, G_2, ..., G_k$, where k is the maximum number of disjoint sets. A vertex s_i in G, is presented $S_1 i, S_2 i, ..., s_k i$ in $G_1, G_2, ..., G_k$. A source node S n the flow network is constructed. For each s_i in S, a vertex $s_o i$ is created to connect s with an edge with the capacity equals to the degree of s_i in G. Also, edges connecting $s_o i$ with $s_j i$ are constructed with the capacity equals to the degree of s_i in G. A vertex X_i is is created to connect every vertex t_{i_j} in G_i with a capacity 1. Two ends Y_1 and Y_2 are created in the flow network. Each vertex X_i is connected to Y_2 with a capacity m. Every vertex t_{i_j} is connected to Y_1 with a capacity n. The *DSC* problem turns to a maximum-flow problem that is to maximize the flow received in Y_2. Finally, the *DSC* problem was formulated and computed by using the mixed integer programming (*MIP*) heuristic.

We considered not only the coverage optimization, but also the connectivity issue. We proposed the Multiple Disjoint Sets with Maintaining Coverage and Connectivity (*MDS-MCC*) problem [13] that given a wireless network with w sinks (or base stations) and n nodes each of which has its respective transmission range and sensing range, and m targets in territory, determine a maximum number of disjoint sets of nodes such that (1) all nodes of each set together cover the whole m targets; (2) for every node in each set, it is connected to a sink via nodes within the same set.

We presented two graph-based models to formulate the *MDS-MCC* problem.

Model 1:*Given integers n, m and w, a directed graph $G = (V, E; f)$, where $R = \{v_{n+1}, ..., v_{n+w} \subset V\}$ is a set of sinks, and $T = \{t_1, t_2, ..., t_m\}$ is a set of targets. Find the maximum integer k such that there exist pairwise disjoint subgraphs $H_1, H_2, ..., H_k$ of $G \backslash R$, and for each supergraph $H_i'(V_i', E_i')$ of H_i, where $V_i' = V(H_i) \cup R$ and $E_i' = \{(v_i, v_j) : (v_i, v_j) \in E, v_i, v_j \in V_i'\}$, satisfying (1) $f(V_i') = T$; (2) for each $v_j in V(H_i)$, v_j is connected to a vertex $u \in R$. $f(v)$ is a labeling function that denotes the set of targets node covers.*

$$f(v_i) = \begin{cases} \{t_j : t_j \in s_i\}, i \leq n \\ \emptyset, i > n \end{cases}$$

And $f(V') = \cup_{v \in V'} f(v)$ where $V' \subseteq V$, $f(H) = \cup_{v \in V(H)} f(v)$ where $H \subseteq G$.

Model 2: *Given integers n, m and w, a directed graph $G = (V, E)$, where $R = \{v_{n+1}, ..., v_{n+w}\} \subset V$ is a set of sinks, and $T = \{t_1, t_2, ..., t_m\}$ is a set of targets. Find the maximum integer k such that there exist pairwise disjoint subgraphs $H_1, H_2, ..., H_k$ of $G \backslash (R \cup T)$, and for each $t_j \in T, t_j$ is connected to u in each supergraph $H_i'(V_i', E_i')$ of H_i, where $V_i' = V(H_i) \cup R \cup T, E_i' = \{(v_i, v_j) : (v_i, v_j) \in E, v_i, v_j \in V_i'\}$ and $u \in R$.*

All sinks can be reduced to only one node r called root by using Lemma 4. Therefore, given either the graph $G = (V, E; f)$ based on Model 1 or the graph $G = (V, E)$ based on Model 2, the corresponding reduced graph can be constructed as $G_r = (V_r, E_r; f, r)$ or $G_r = (V_r, E_r; r)$ where $V_r = \{v_1, v_2, ..., v_n, r\}$, the first n vertex are the nodes, and r is root. There is no more changes expect that an edge in E from a node to a sink becomes an edge in E_r from the node to the root. If a node is connected to more than one sink, only one edge is added in G_r.

Lemma 4.[13]*A node is connected to a sink if and only if it is connected to the root in the corresponding reduced graph.*

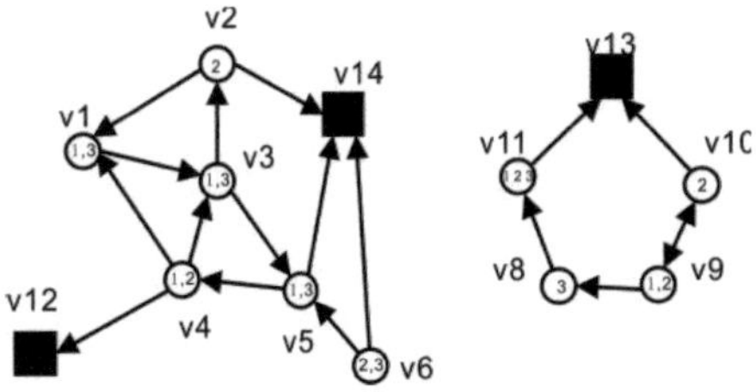

(a) Original Network Topology Graph G

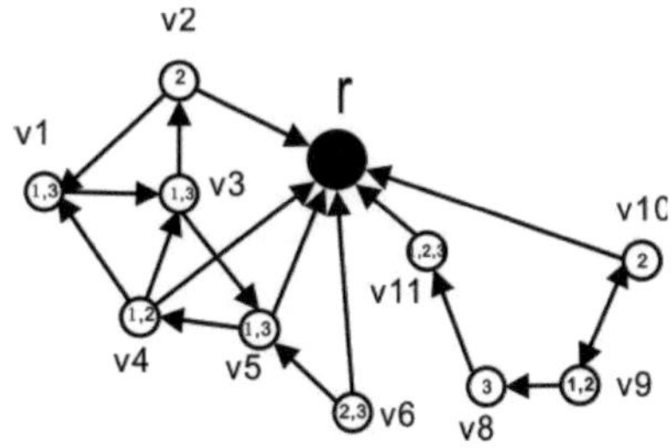

(b) Reduced Graph G_r of G

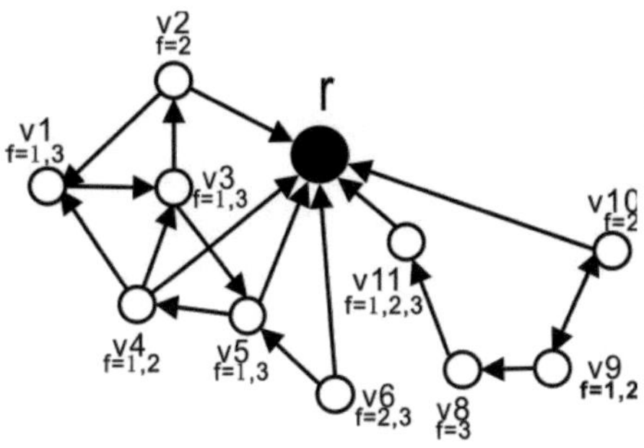

(c) Constructed Graph based on Model 1

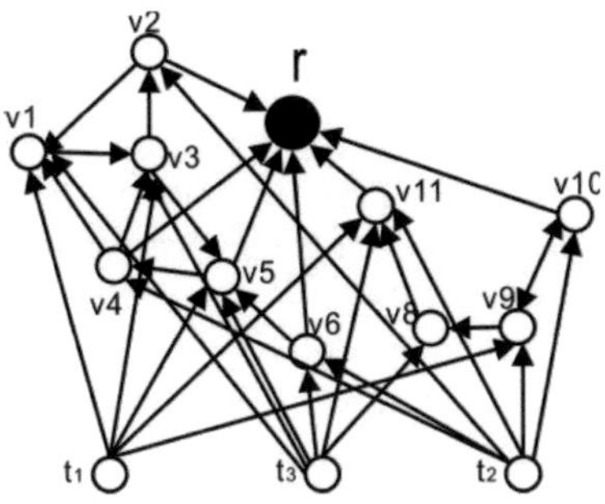

(d) Constructed Graph based on Model 2

Figure 3. Example of original network topology (a) and its corresponding reduced graph (b). (c) and (d) shows the constructed graph based on the Model 1 and Model 2 respectively.

We found that *MDS-MCC* problem is *NP-complete*. Two algorithms, heuristic and network flow were proposed to solve *MDS-MCC* based on the two models. Heuristic algorithm was designed to find the maximal number of disjoint subgraphs of G based on Model 1. The algorithm first initializes that $H = \emptyset$, which presents the set of vertex having been found. Then the algorithm finds out all the paths whose ends belongs to H and put the vertex in these paths into H until H covers all of the targets. The algorithm deletes the redundant nodes from H when H still covers all of the targets after delete it. The algorithm finds one subgraph H and repeats to find other subgraphs.

Algorithm 5.[15]*Heuristic Algorithm for MDS-MCC.*
1:Construct reduced graph $G_r = (V_r, E_r; f, r)$ *from* $G = (V, E; f)$ *and T;*
2:k=0;
*3:***WHILE***still has the subgraph of G that covers all the targets* **DO**
4: $k = k + 1$;
5: $H_k = \{r\}$;
6: **WHILE** $|f(H_k)|$ łm **DO**
7: $PS = \{v_0 v_1 ... v_e : v_e \in H_k, v_i \in V_r - H_k, f(v_j) \subseteq f(H_k), f(v_0) \notin f(H_K)\}$;
8: *Select one path* $P \in PS$;
9: $H_k = P \cup H_k$;
10: ***END WHILE***
11: *Delete redundant nodes from* H_k;
12: $G_r = G_r \backslash H_k$;

http://dx.doi.org/10.5772/54146

*13:**END WHILE***
*14:**RETURN*** $H_1, H_2, ..., H_k$

In Model 2, the problem is converted to find the maximal k disjoint sets such that there exists a path from every target node t to the root node r for each set. For any two paths from a target node t to the root r belonging to any two disjoint sets, the two paths are disjoint. Therefore, there are k pairwise disjoint paths from t to r.

The network flow algorithm first finds a set of paths $P_i = \{p_i^1, p_i^2, p_i^3, ..., p_{i'}^i\}$ for every target node t_i to r such that any two paths are disjoint. We converted the problem of obtaining the related paths with maximum l_i from t_i to r to the maximum flow problem like *DSC*. We presented a method to construct the flow network from the network of Model 2. An example is shown in Fig.4. The problem that finds the maximum number of paths is converted to solve the problem of computing the maximum flow from s to Y_2. For each copy in the flow network, the flow network algorithm chooses the vertex for each disjoint result set if the flow from the vertex v_0j to the vertex v_ij is greater than 0.

Algorithm 6.[13]*Network Flow Algorithm for MDS-MCC.*
1:Construct $G_r = (V_r, E_r; r)$ *from* $(G = (V, E))$ *and* T;
*2:**FOR EACH*** $t_i \in T$ ***DO***
3: Find l_i *pairwise disjoint paths,* $p_i^1, p_i^2, ..., p_{i'}^i$ *from* t_i *to* r;
*4:**END FOR***
5:Find the maximum k pairwise disjoint sets $H_1, H_2, ..., H_k$ *such that each set* $H_i = \cup_{j=1}^{m}\{V(P_j^{i_j}) - \{t_j, r\}\}$;
*6:**RETURN*** $H_1, H_2, ..., H_k$;

A special case was studied in wireless sensor networks that each node covers at most one target. We assumed there are m targets that each is exactly monitored by k sensor nodes. There must be k disjoint sets each of which completely covers all the targets and is connected to one of sinks. And k is the theoretical maximum number.

Theorem 12.[14]*Given integers n and m, a directed graph* $G = (V, E; f, r)$ *and a target set* $T = \{t_1, t_2, ..., t_m\}$.$A_1, A_2, ..., A_m$ *be m disjoint sets with* $|A_i| = k$, *where* $A_i \subset V(G)$. *If* $k = 2$ *and G is* $(m + max\{1, m-4\})$*-connected, or* $k \geq 3$ *and G is* $(m(k-1)+1)$*-connected, then there exist k connected subgraphs* $H_1, H_2, ..., H_k$, *satisfying (1)* $f(H_i) = T$; *(2) for each* $v_j \in V(H_i), v_j$ *is connected to* r.

5. Minimum Spanning Subgraph and Minimum Steiner Tree

The minimum spanning subgraph is used to reduce the cost of algorithms in underlying wireless ad-hoc networks that are modeled as graphs. For instance, a spanning tree is used as a backbone to reduce the cost of broadcast, or to cluster the hierarchical structure. A spanning tree of a graph is a subgraph that is a tree and connects all the vertices together. Many research works on ad hoc related networks exploited minimum spanning subgraph to design energy-efficient distributed protocols, multicast routing protocols, fault tolerant topology control protocols and etc. To construct spanning subgraph has the advantage of low time and message complexity.

The Steiner tree problem is superficially similar to the minimum spanning tree problem. The minimum Steiner tree problem is a problem in combinatorial optimization, which may

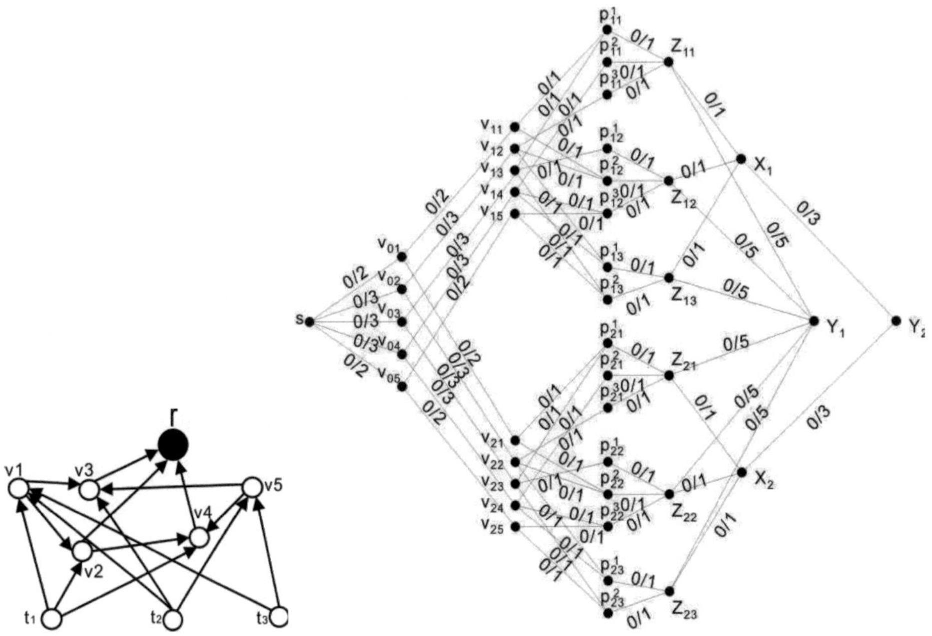

(a) Network Topology

(b) Constructed Flow Network

Figure 4. [13] Example of converting the network topology based on Model 2 to a constructed flow network.

be formulated in a number of settings, with the common part being that it is required to find the shortest interconnect for a given set of objects. Most of the Steiner Tree problems are NP-complete. Similar with spanning tree, Minimum Steiner Tree problem has been applied to design minimum energy broadcasting protocols, delay constrained protocols, range assignment topology control protocols in ad hoc networks.

5.1. Minimum Spanning Tree and Subgraph

Given a graph $G(V, E; C)$, a spanning tree of G is a subgraph that is a tree and connects all the vertices together. A minimum spanning tree (*MST*) is the spanning tree with the minimum total weight in all spanning trees. There are many extended concepts of *MST*. The k-minimum spanning tree (k-*MST*) is the tree that spans some subset of k vertices in the graph with minimum weight. A set of k-smallest spanning trees is a subset of k spanning trees such that no spanning tree outside the subset has smaller weight. The Euclidean minimum spanning tree is a spanning tree of a graph with edge weights corresponding to the Euclidean distance between vertices that are points in the plane. The degree constrained minimum spanning tree is a minimum spanning tree with each vertex v is connected to no more than $d(v)$ other vertices. If the graph G is not connected, the graph has minimum spanning forest. Similarly, a minimum spanning subgraph is the spanning subgraph with the minimum total weight in all spanning subgraphs.

M. Cardei et al. [16] addressed the topology control issue of power assignment in ad-hoc networks by using minimum spanning subgraph techniques. They aimed to minimize the total transmission power assigned for all nodes while building k-vertex fault tolerant communication paths from each node to the sinks. They modeled the network topology with an undirected weighted graph. The issue was formulated as that given a directed graph G and a root r, find a directed spanning subgraph of G such that: (1) the sum of the weight of the edges is minimized, (2) there are k-vertex disjoint paths between r and every vertex in G. The *FT* algorithm [18] was applied to solve this problem optimally to reach k-approximate.

We addressed the issue of k-vertex fault-tolerant many-to-many routing power assignments in ad-hoc networks [17] that given an ad-hoc network consisting of n nodes with the various transmission ranges. For each node V_i, it can adjust the transmission ranges up to its maximum value R_max. Determine the power p_i of node V_i such that 1) there exist k-vertex disjoint data routing paths between any pair of nodes; 2) the total power consumed over all sensor nodes is minimized, namely $\sum_{i=1}^{n} p_i$ is minimized.

A directed weighted graph $G(V, E; C)$ was represented to model the network topology, where $V = \{v_1, v_2, ..., v_n\}$ is the set of nodes and $E = \{(v_i, v_j) : dist(v_i, v_j) \leq R_max\}$ is the set of edges. For each edge $(u, v) \in E$, there exists a weight $C(u, v)$ associated with it. $C(u, v)$ represents the power consumption needed by u to communicate with v. It aimed to construct a minimum k-vertex connected subgraph of G by finding a set of power assignments for each node.

Two algorithms were proposed to find such minimum k-vertex connected subgraph of G by using spanning subgraph technology. The first algorithm produces a k-vertex connected spanning subgraph and assigns to each vertex the minimum transmission range to reach all of its neighbors. The algorithm removes the edges in decreasing order of their weights if and only if the graph keeps k-vertex connected after the removal. Theorem 13 guarantees that the final remaining subgraph is k-vertex connected. The algorithm assigns the transmission power to each node according to the subgraph.

Theorem 13.[17] *A graph $G(V, E)$ is a k-vertex connected directed graph. If $(u, v) \in E$ and there are at least $k + 1$ disjoint paths from u to v, namely $\lambda(u, v) = k + 1, G - (u, v)$ is a k-vertex connected graph.*

Algorithm 7.[17] *Heuristic Algorithm for k-vertex fault tolerant power assignments.*
1:Sort all edges in E in decreasing order of the weights;
*2:**FOR EACH** edge (u, v) in the sorted order **DO***
*3: **IF** $\lambda(u, v) = k + 1$ **THEN***
4: $G = G - (u, v)$;
*5: **END IF***
*6:**END FOR***
*7:**FOR** $i = 1$ **TO** n **DO***
8: $p_i = max\{C(v_i, v_j) : (v_i, v_j) \in E\}$;
*9:**END FOR***

Another algorithm is an $O(\sqrt{n/\varepsilon})$-approximation algorithm [17] by using the solution of the minimum-cost k-vertex connected spanning subgraph problem proposed by Cheriyan k-vertex connected, for any $\varepsilon > 0$ and $k \leq (1 - \varepsilon)n$.

5.2. Minimum Steiner Tree

Steiner tree problem (*STP*) is a classical combinatorial optimization problem. This problem has a lot of versions. The graph version of *STP* is that: Given a edge-weighted graph $G =$

$(V, E; C)$ with edge-weight function $C : E \rightarrow \Re^+$ and a subset $U \subset V$ called a terminal set, the *STP* is to find a subtree T of G interconnecting the terminal set U with minimum total weight. The graph version of *STP* is *NP-complete* and the best approximation ratio is $\rho = 1 + \frac{\ln 3}{2} = 1.55$ till now [28].

We considered a variant of *STP* – Node-Weighted Steiner Tree (*NWST*) problem, that is, the weight function is from vertex set to positive real set now. Given a node-weighted graph $G = (V, E; C)$ with node-weight function $C : V \rightarrow \Re^+$ and a subset $U \subset V$ called a terminal set, the *STP* is to find a subtree T of G interconnecting the terminal set U with minimum total weight. As an application, *NWST* can be used on *NWCDS* problem to interconnect the node-weighted dominating set.

To deal with the *NWST* problem, the first idea is to convert this problem to classical *STP*. We constructed a new graph G' with the same vertex set, edge set and terminal set. The difference is the weighted function of G' is on edges.

Algorithm 8.[9]*Approximation Algorithm for NWST on UDGs*
1:*Construct an edge-weighted graph* $G' = (V, E; C')$ *with the same vertex set, edge set and terminal set of* G*;*
2:***FOR EACH*** *edge* u, v *in graph* G' ***DO***
3: *Assign the weight of this edge* $C'(u, v) = (C(u) + C(v))/2$*;*
4:***END FOR***
5:$T = SMT(G', U)$*, where* $SMT(G', U)$ *is the best-known approximation algorithm on graph* G' *and terminal* U*;*
6:***RETURN*** T*;*

Theorem 14.[9]*Algorithm 8 is a* 2.5ρ*-approximation for node-weighted Steiner tree problem in unit disk graph.*

Furthermore, we should give a theoretical result, to show that the *NWST* has polynomial-time approximation scheme (*PTAS*) on *UDGs* if the terminal set U is c-local, that is, in the minimum node-weighted spanning tree for U, the Euclid distance of the longest edge is at most some constant c. A *PTAS* is a family of approximation algorithm with ration $1 + \varepsilon$ for any $\varepsilon > 0$.

The main idea of the *PTAS* for *NWST* is based on the partition and shifting strategy. Firstly, we partitioned the whole area containing all vertices into some small cells. Then, we divided every cell into interior area and boundary area. Secondly, for each cell, we constructed a local optimal Steiner forest on terminal vertices in the interior area of this cell. Then, we combined all these forests to obtain a local optimal Steiner forest. Thirdly, we added all the crossing edges to get a Steiner tree on terminal set R. We call the resulting graph G_p for a specific partition P. In order to get a better node-weighted Steiner tree, we shifted the partition and choose the minimum output among all of partitions.

Theorem 15.[6]*Node-weighted Steiner tree problem has PTAS on unit disk graphs.*

6. Conclusion

Many problems in mobile ad-hoc networks can be formulated by using graph theory. However, in real ad-hoc network applications, there are many constraints that make the

http://dx.doi.org/10.5772/54146

issues difficult to be tractable. The methods that might convert these practical issues into graph-based problems are important for the design and implementation of routing, broadcasting and organization algorithms. In this chapter, we present three essential graph-based issues casted from practicle ad-hoc network issues: Connected Dominating Set, Disjoint Sets, and Minimum Spanning Subgraph and Minimum Steiner Tree. Theoritical analysis are described to verify the correctness of these proposed algorithms that are either heuristics or approximationg.

While much efforts have been made to solve the graph-based issues, still much progress needs to be done. For instance, some clustering issues in ad-hoc networks can be casted into graph labeling or graph coloring problem that assigns of labels to the nodes subject to certain constraints. In other mobile ad-hoc applications, many problems involve graph classification, graph subsumption, and even the description and implementation of graph data structure, querying and database. More approaches for ad-hoc network applications should be discussed from the aspectes of the applicability and the utility by using graph theory.

Acknowledgements

This work was partially supported by the Scientific Research Foundation for the Returned Overseas Chinese Scholars, State Education Ministry, the Gansu Provincial Science & Technology Department (grant no. 1007RJYA010), the National Natural Science Foundation of China (grant nos. 61003240, 11201208, 10905026, 11275003) and the Fundamental Research Funds for the Central Universities (grant no. lzujbky-2011-44).

Author details

Li Liu[1], Xianyue Li[2], Jiong Jin[3], Zigang Huang[4],
Ming Liu[5] and Marimuthu Palaniswami[3]

1 School of Information Science and Engineering, Lanzhou University, P.R.China
2 School of Mathematics and Statistics, Lanzhou University, P.R.China
3 Department of Electrical and Electronic Engineering, The University of Melbourne, Australia
4 School of Physical Science and Technology, Lanzhou University, P.R.China
5 School of Electrical and Information Engineering, The University of Sydney, Australia

References

[1] B.N. Clark, C.J. Colbourn and D.S. Johnson, Unit Disk Graphs, Discrete Mathematics, 86 165-177 (1990).

[2] X. Gao, Y. Wang, X. Li and W. Wu, Analysis on Theoretical Bounds for Approximating Dominating Set Problems, Discrete Mathematics, Algorithms and Applications, 1 71-84 (2009).

[3] X. Gao, X. Li and W. Wu, A Constant-Factor Approximation for d-Hop Connected Dominating Set in Unit Disk Graph, The 10th International Conference on Information and Management Sciences (IMS 2011), 263-272 (2011).

[4] D. Kim, W. Wang, X. Li, Z. Zhang and W. Wu, A New Constant Factor Approximation for Computing 3-Connected m-Dominating Sets in Homogeneous Wireless Networks, The 29th IEEE Conference on Computer Communications (INFOCOM 2010), 1-9 (2010).

[5] D. Kim, Z. Zhang, X. Li, W. Wang, W. Wu and D.-Z. Du, A Better Approximation Algorithm for Computing Connected Dominating Sets in Unit Ball Graphs, IEEE Transactions On Mobile Computing, 9 1108-1118 (2010).

[6] X. Li, X.-H. Xu, F. Zou, H. Du, P.-J. Wan, Y. Wang and W. Wu, A PTAS for Node-Weighted Steiner Tree in Unit Disk Graphs, The 3rd Annual International Conference on Combinatorial Optimization and Applications (COCOA 2009), 36-48 (2009).

[7] M. Min, H.-W.Du, X.-H. Jia, C.-X. Huang, S.-C. Huang and W. Wu, Improving Construction for Connected Dominating Set with Steiner Tree in Wireless Sensor Networks, Journal of Global Optimization, 35 111-119 (2006).

[8] W. Shang, F. Yao, P. Wan and X. Hu, On Minimum m-Connected k-Dominating Set Problem in Unit Disc Graphs, Journal of Combinatorial Optimization, 16 99-106 (2008).

[9] F. Zou, X. Li, S. Gao and W. Wu, Node-weighted Steiner tree approximation in unit disk graphs, Journal of Combinatorial Optimization, 18 342-349 (2009).

[10] F. Zou, X. Li, D. Kim and W. Wu, Construction of Minimum Connected Dominating Set in 3-Dimensional Wireless Network, The 3rd International Conference on Wireless Algorithms, Systems and Applications (WASA 2008), 134-140 (2008).

[11] F. Zou, Y. Wang, X.-H. Xu, X. Li, H. Du, P. Wan and W. Wu, New approximations for minimum-weighted dominating set and minimum-weighted connected dominating sets on unit disk graphs, Theoretical Computer Science, 412 198-208 (2011).

[12] M. Cardei, D.-Z. Du, Improving wireless sensor network lifetime through power aware organization, Wireless Networks, 11(3) 333-340 (2005).

[13] L. Liu, B. Hu, L. Li, Energy Conservation Algorithms for Maintaining Coverage and Connectivity in Wireless Sensor Networks, IET Communications, 4(7) 786-800 (2010).

[14] H. Li, H. Miao, L. Liu, L. Li, H. Zhang, Energy conservation in wireless sensor networks and connectivity of graphs, Theoretical Computer Science, 393(1-3) 81-89 (2008).

[15] L. Liu, B. Hu, H.F. Miao, H. Li, L. Li, Q.L. Zhao, Achieving Energy Conservation, Coverage and Connectivity Requirements in Wireless Sensor Networks, The 29th IEEE International Conference on Distributed Computing Systems Workshops (ICDCS Workshop 2009), 227-232 (2009).

[16] M. Cardei, S. Yang, J. WU, Algorithms for fault-tolerant topology in heterogeneous wireless sensor networks, IEEE Transaction on Parallel Distributed Systems, 19(3) 545-558 (2008).

[17] L. Liu, L. Li, B. Hu, Algorithms for k-fault Tolerant Power Assignments in Wireless Sensor Networks, Science China-Information Sciences, Springer, 53(12) 2527-2537 (2010).

[18] N. Li, J.-C. Hou, FLSS: A fault-tolerant topology control algorithm for wireless networks, The 10th Annual International Conference on Mobile Computing and Networking (MobiCom 2004), 275ĺC286 (2004).

[19] J. Cheriyan, S. Vempala, A. Vetta, Approximation algorithms for minimum-cost k-vertex connected subgraphs, The 34th Annual ACM Symposium on the Theory of Computing (STOC 2002), 306-312 (2002).

[20] D. Chakrabarti, C. Faloutsos, Graph mining: laws, generators, and algorithms, ACM Computing Surveys, 38(1) (2006).

[21] A.-L. Barabási, Linked: the new science of networks, Perseus Books Group (2002).

[22] L. Liu, B. Hu, L. Li, Algorithms for Energy Efficient Mobile Object Tracking in Wireless Sensor Networks, Cluster Computing, 13(2) 181-197 (2010).

[23] J. Tian, J. Hähner, C. Becker, I. Stepanov, K. Rothermel, Graph-based mobility model for mobile ad hoc network simulation, The 35th Annual Simulation Symposium, (2002).

[24] A. Casteigts, S. Chaumette, Dynamicity aware graph relabeling systems (da-grs), a local computation based model to describe manet algorithms, The 17th International Conference on Parallel and Distributed Computing and Systems (PDCS 2005), 231-236 (2005).

[25] S. Bittner, W.-U. Raffel, and M. Scholz, The area graph-based mobility model and its impact on data dissemination, The 3rd IEEE International Conference on Pervasive Computing and Communications Workshops (PerCom Workshops 2005), 268-272 (2005).

[26] M. Jahnke, C. Thul, P. Martini, Graph based metrics for intrusion response measures in computer networks, The 32nd IEEE Conference on Local Computer Networks (LCN 2007), 1035-1042 (2007).

[27] W. Ke, P. Basu, T.D.C. Little, A task graph based application framework for mobile ad hoc networks, The 5th IEEE International Conference on Communications (ICC 2002), 3279-3283 (2002).

[28] G. Robins and A. Zelikovsky, Improved Steiner tree approximation in graphs, The 11th Annual ACM-SIAM Symposium on Discrete Algorithms, SODA, 770?779 (2000).

Chapter 4

Probabilistic Routing in Opportunistic Ad Hoc Networks

Vangelis Angelakis, Niki Gazoni and Di Yuan

Additional information is available at the end of the chapter

http://dx.doi.org/10.5772/53997

1. Introduction

Routing packets in multi-hop ad hoc wireless networks poses a great challenge mainly due to the unreliability of the wireless links and the inherent interference of the wireless medium [13]. Due to these characteristics, traditional wired routing schemes that select the best path towards a destination and forward the packet to a specific next hop, have proven ill-suited for networks utilizing the wireless medium and relying on lossy broadcast links. Lately, a new routing paradigm, namely *opportunistic routing*, has been proposed to cope with unreliable transmissions by taking advantage of the broadcast nature and spatial diversity of the wireless medium [2, 6].

Opportunistic routing constitutes a new routing paradigm leveraging the nodes' ability to overhear a broadcast packet. Its core difference from traditional routing schemes in that forwarders can be selected from the group of recipients of the packet even after its transmission, hence there is no hard commitment to a predetermined path. This flexibility enables opportunistic routing to combine multiple weak links to create a reliable route, as well as to exploit unexpectedly long transmissions. The added forwarding reliability in transmission reduces the retransmission cost, which in turn improves the throughput and energy efficiency [10, 16]

Proposed opportunistic protocols demonstrate a lack of concrete understanding of the way key wireless networking primitives and design decisions affect the performance of an opportunistic routing scheme. As a result, it is unclear to which extent the improved performance of these protocols owes to their opportunistic design and to which extent it is affected by other design features that can also be applied to traditional routing schemes [19, 20, 22]. Opportunistic protocols which decide on forwarders in a centralized manner require the exchange of node coordination messages, leading to high overhead and increased resource consumption. Furthermore they require global knowledge of the topology which makes them prone to poor performance in the event of misinformation.

On the other hand, localized probabilistic forwarding decision protocols, which have been designed mostly for use in sensor networks, have to trade high performance for robustness and simplicity. In fact some of them partly rely on flooding techniques, thus demonstrating a percolation behavior, which leads to unnecessary transmissions [23].

In designing of an opportunistic routing scheme, there are two key design decisions [10, 11]. Firstly, how node receiving a packet should decide to forward or not. Secondly, provided that a node has decided to relay a packet, based on some metric, when is the most appropriate time to do so. Here, using a simulation framework, we examine how the forwarding decisions and transmission timing affect performance and under which channel error conditions and topology density it is beneficial to use opportunistic routing instead of traditional routing. Furthermore, we develop an opportunistic forwarding scheme, whose parameters can be tuned to allow for low resource consumption and high delay performance, while being robust to misinformation. We provide evidence which confirms that the suggested protocol supports multiple flows in a network, a weakness of existing early demonstrated solutions in the literature. Finally, to evaluate the scheme, and the role of the wireless primitives in forwarding, we provide comparisons to single-path routing and two opportunistic routing protocols, SOAR [15] and Directed Transmission [12]. Our simulation results, under various channel error and misinformation conditions, demonstrate that the proposed routing protocol outperforms both SOAR which uses a centralized scheme for forwarding decision-making and Directed Transmission which, designed for sensor networks, is distributed using only local information.

1.1. Routing schemes in multi-hop wireless networks

1.1.1. Single-path routing

Initially, the routing techniques that have traditionally been used in wired networks were transported for use in multi-hop wireless networks as well. These routing protocols typically rely on choosing the best sequence of nodes between a source and destination, by some metric, and then forward each packet through that path, until something goes wrong, i.e. packets start getting lost. However, in the wireless domain, the performance of single path routing would often prove unsatisfactory due to the fact that, especially in multi-hop networks, link conditions are highly varying due to interference that such a single path may be far from optimal. Typically such protocols yield highly volatile routes incurring very high overhead costs in end-to-end communication. Most of the well-established routing protocols, such as DSR [7], AODV [14], and DSDV [4] fall into this category.

1.1.2. Multi-path routing

Towards improving routing performance, in multi-hop wireless networks, multi-path routing takes advantage of the potential of multiple alternative paths between a source-destina-

http://dx.doi.org/10.5772/53997

tion pair. The multiple paths established can be overlapping, edge-disjoint or node-disjoint with each other [6]. Data traffic is then split, according to some routing goal, along these multiple paths avoiding congestion and to using network resources with increased efficiently. Multi-path routing can yield a variety of benefits such as fault tolerance, increased bandwidth, or improved security. It is typically proposed in order to increase the reliability of data transmission or to provide load balancing [11].

1.1.3. *Opportunistic routing*

Opportunistic routing differs from the above traditional routing techniques in that it leverages the broadcast nature of wireless medium and the route is typically generated on the fly, i.e. the transmitting nodes may vary dependent on the actual packet broadcast transmission. As opposed to multi-path routing, it can remain "commitment-free" to the number of predetermined paths. Among the nodes that receive the packet, the one with the best conditions to the destination may be chosen to relay [3, 17, 21]. This can mitigate the effect of losses or increased overhead due to unreliable and unpredictable wireless links.

The key opportunistic routing benefits are the following two [15]:

i. Opportunistic routing can be used to combine multiple weak links to produce a virtual strong link.

As an example consider a node with five candidate relay neighbors and a distant destination and assume there is a 0.2 packet success rate to each of the neighbors, while each of them has a packet success rate of 1.0 to the destination. Under a traditional routing protocol, one of the five neighboring nodes would be selected as relay, resulting, on average, in five transmissions to send a packet from the source to the relay node, and then one more transmission from the relay node to the destination.

Opportunistic routing can consider the five neighboring nodes as one virtual link that can forward the packet to the destination. This opportunistic virtual link will then have a success rate of: $1 - (1 - 0.2)^5 = 0.672$. So, on average only 1/0.67=1.487 transmissions will be required to deliver a packet to *at least one* of the five intermediate nodes, and one more transmission is required for an intermediate node to forward. Hence, with 2.487 transmissions required on average end-to-end packet delivery through the opportunistic virtual link, achieves a throughput gain of x2.4 over traditional routing.

ii. Opportunistic routing can take advantage of "against-the-odds" successful transmissions, to achieve increased throughput.

In traditional routing protocols there typically is a trade-off reflected in the route selection metric between link quality and the spatial progress, in terms of distance from the destination, each transmission achieves. Consider a tandem of 4 nodes A, B, C, and D where the channel conditions from A to B is expected to be better than those of A to C, and so on due to proximity. Then indeed a transmission from A to B is more likely to be successful, albeit the information will physically progress less hence an added transmission to C may be required to guarantee, with some probability margin, reaching the final destination.

Not, in advance, committing to who will forward a packet to the final destination in such a case can prove beneficial if for some reason the channel conditions prove unexpectedly favorable to the long transmission from A to C. Hence a rule of thumb stating that: "Among the nodes that receive the packet, the one closest to the destination should forward" comes naturally in the case of opportunistic routing.

1.2. Related opportunistic routing works

Here, we introduce the major opportunistic routing protocols our work is based on, and/or is compared against.

1.2.1. Extreme Opportunistic Routing: ExOR

ExOR [1] is a routing protocol for wireless multi-hop networks that was implemented on the RoofNet testbed at the Massachusetts Institute of Technology. ExOR integrates routing and MAC protocols, still all packets are broadcast at layer 3. It improves routing performance by utilizing opportunistically success over unstable long-range links (see above).

After collecting a batch of packets, which is uniquely identified by a *BatchID*, the source broadcasts each packet in the batch, listing the forwarding nodes in a priority order in the packet's header. Priority is set by the ETX metric to the destination [8]. The ETX metric, i.e. essentially the expected number of transmissions necessary to forward a packet along a route, can be considered to map to a distance from the destination, hence, the lower the metric, the higher the priority. Only nodes that are "closer" to the destination than the source are included in the potential forwarder set. Each packet carries a bitmap, marking packets that have been received by the sending node or nodes with higher priorities. A forwarder transmits a packet only if no forwarder with higher priority has explicitly acknowledged receipt of it. ExOR has good routing performance.

It is unlikely that a forwarder will receive the entire batch correctly, so the nodes that have stored fragments of the batch will need to schedule their transmissions. To that end, each forwarder uses a forwarding timer that ExOR has set to five packet durations times the number of higher priority nodes in the forwarder list, and which is a prediction of the time at which the node should start forwarding packets from its packet buffer. After each schedule cycle, the batch maps need to be updated by means of negative acknowledgments and when the destination has received 90% of the batch, the rest of it is sent using traditional routing, because the overhead would be forbidding otherwise.

Due to the centralized coordination and scheduling that is needed between forwarders and the destination, ExOR incurs high overhead when the batch of packets to transmit is small as in bursty and short-lived flows, or the number of candidate forwarders is large.

1.2.2. The Simple Opportunistic Adaptive Routing protocol: SOAR

The SOAR protocol [15] has been proposed as an improvement to ExOR in order to support multiple flows. In SOAR the candidate forwarders are constrained to be on-or-near the shortest path from source to destination. The cost metric is again ETX. A significant difference between ExOR and SOAR is that SOAR performs the routing decision process on a *per-packet* basis rather than on a batch. Finally, in SOAR the forwarder list is limited to 5 relays. Overall, SOAR incurs slightly less overhead than ExOR and restricts flows as close to the shortest path from source to destination as possible. It is considered crucial to avoid diverging paths, and select forwarders that can hear each other because overhearing is the only means to avoid duplicate transmissions.

To make the protocol reliable to ACK losses, selective ACKs are used. Each ACK packet contains the starting sequence number of out-of-order ACKs and a bit-map of out-of-order ACKs. In addition, SOAR uses a scheme of *piggy-backed acknowledgement* and *ACKs compression* to reduce the overhead of acknowledgments. When a node does not have much data to send, it should also send stand-alone ACKs to provide timely channel feedback.

However, the increase in throughput that is observed when SOAR is in use is partly due to its opportunistic scheme and partly due to its complex acknowledgment scheme, therefore making it unclear at what extent this protocol contributes to opportunistic routing [20].

1.2.3. Resilient Opportunistic Mesh Routing: ROMER

In ROMER [18], a credit mechanism is used to build a forwarding mesh topology on the fly centered on minimum cost paths. It aims to avoid the cost of retransmissions over persistently poor routes. Each packet is assigned a number of credits equal to the amount needed to reach the destination along the minimum cost path plus some extra slack allowing the packet to deviate slightly from this (shortest) path

Before forwarding a packet the transmitting node calculates the remaining credit that the packet would have if forwarded over a link and compares it to a threshold value. This scheme enables multiple forwarders to emerge near the source to ensuring route diversity and packet progress in view of lossy links. At the same time the scheme secures converge near the destination and remains close to the minimum cost path.

Greedy probabilistic forwarding is used to deliver the data packet along the instantaneously highest rate link with probability one and along other high-rated downstream links with a high probability value. Thus, it favors high quality links over low quality ones by selecting higher random forwarding probability. However, it does not account for resource conservation and catering to multiple flows.

1.2.4. Destination attractor and directed transmission

The Parametric probabilistic sensor network routing scheme of [15] proposes two protocols that forward a single packet with varying retransmission probability through a network of sensor nodes, focusing on simplicity and robustness to errors in distance estimation. The Destination Attractor assigns a higher retransmission probability to the packet, as it moves closer to the destination and reduces it, as the packet moves away from the destination. Distance check is performed by comparing the distance of the source node in hops from the destination to the distance of the node currently holding the packet. The primary concern of this approach is to deliver as many copies of a single packet to the destination sensor as possible, without accounting for resource usage or delays. Directed Transmission improves probabilistic routing's performance by assigning an exponentially higher forwarding probability to nodes that are on the shortest path from the source to the destination and decreasing it as the packet strays from the shortest path. This leads to lower resource consumption than the Destination Attractor and can be tuned to resemble shortest path routing, when the misinformation is low enough. Both protocols are compared to shortest path routing as an ideal case.

2. Design considerations for a probabilistic routing scheme

As it has become apparent from our discussion thus far, in designing a probabilistic packet routing scheme for an opportunistic ad hoc network there are two key design decisions in the core of the forwarding procedure [5]:

a. *how* a node should decide whether or not to forward a packet it received and is not intended for him, and

b. *when* is the most appropriate time to transmit such a packet.

In what follows we develop a concise simulation framework to investigate how the forwarding decisions and transmission timing affect performance, and under which channel error conditions and topology density it is beneficial to use stochastic opportunistic routing instead of traditional routing techniques. With the insight gained, we develop a probabilistic packet forwarding procedure can be automatically tuned to allow for low resource consumption and delay-tolerant performance, while at the same time being robust to poor system information, which is a common case in ad hoc opportunistic networks. The procedure is embedded in a basic routing scheme, which is compared to single path routing and two reference opportunistic routing protocols, already discussed, SOAR and Directed Transmission. Simulation results under various channel error and errors in the metric calculation, demonstrate that our proposed protocol outperforms both SOAR, which uses a centralized forwarding decision scheme and Directed Transmission, which is highly distributed and designed for sensor networks.

2.1. Forwarding procedure principles & parameters

Apart from an initial neighbor discovery phase, routing decisions should require minimal information to be exchanged between nodes and no co-operation, to avoid imposing computational load on nodes with limited computational capacity and wasting bandwidth in exchanging control packets when bandwidth is limited. The forwarding procedure developed therefore had the key goals of simplicity and distributed decision-making. This is an obvious trade-off, as the simplest design solution would be flooding, a technique used in many sensor network opportunistically routing schemes (e.g. see [23] and the references therein).

http://dx.doi.org/10.5772/53997

The immense resource consumption of this scheme renders it completely impractical when amongst the targets of a routing protocol is e.g. to allow for multiple flows to exist simultaneously in the network – a goal not always necessary in event-detection sensor networks. Summarily the stochastic forwarding procedure proposed is intended as a flexible solution in routing applied in a variety of networks, where lossy areas are observed due to low link quality and that the topology is sufficiently dense to require opportunistic communication.

The major issues that need to be addressed in the design of the routing scheme that encompasses a stochastic forwarding procedure are (i) deciding which nodes should forward a packet, (ii) when to do so, (iii) how to acknowledge the reception of a packet. Candidate forwarders should be selected in order to improve the performance of the end-to-end flow. On the other hand the number of these forwarders should be limited to those that guarantee, with a high probability, the packets' progress towards their destination, in order to avoid excess resource consumption and allow better network resources utilization. Furthermore, the forwarding time differentiation between the candidates enables the mitigation of redundant transmissions, by limiting packet collisions. Lastly, since layer 3 broadcast packets typically do not implement link-layer acknowledgements, such a scheme should be devised. The destination could generate acknowledgements upon packet reception, which would be propagated back to the source, either opportunistically or via the shortest path. This would impose significant load on the network if the acknowledgements were "stand-alone" layer 2, or layer 3 control packets competing with the data packets and being susceptible to lossy links, which would cause significant delay until retransmission of a lost packet. However, the option of piggy-backing the acknowledgments onto data packets is limiting since it can be applied only when flows of symmetric direction exist between source and destination. Hop-by-hop acknowledgments provide robustness to the lossy links and mitigate delay on the other hand increase the overhead. Conversely, a passive acknowledgment scheme that utilizes overhearing of other nodes' transmissions, avoids imposing additional load to the network, contrary to explicit acknowledgment packets.

2.2. Forwarding cost metric

For the purpose of determining which nodes are the most suitable forwarding candidates for a particular packet, a routing cost metric needs to be used. For the framework presentation elucidation, the procedure presented here uses as a metric the physical distance between two nodes. However, any cost metric can be used to this end, (provided, the metric has well-defined minimum and maximum values). In order for a node to be able to calculate its routing metric from the destination of the packet, a neighbor discovery phase should take place between the nodes in network, before data packets can be exchanged. In our discussion we consider this to have taken place offline and do not assume it to be a part of the protocol, although it is an aspect that would increase the protocol overhead. Note that for the actual simulation comparisons the same assumption has been made for all protocols. Other metrics' calculation, as for example the ETX metric used in ExOr, can be incorporated in the specific protocol, with relative low design overhead. Note that such processes may be prone to misinformation due to the overly distributed nature of the network.

2.3. Forwarding probability function

In order to determine which nodes will forward a packet in a stochastic manner, a forwarding probability function is utilized. This function is the same for all nodes in the network. It's role is to map cost metrics to forwarding probabilities to each node that receives a packet. More specifically, this function is (a) non-increasing in the cost metric, i.e. lower routing cost values should not yield lower probability, and (b) bound, so that the minimum values it assigns are between zero and one. Given the cost metric assumed, our forwarding probability is assigned with respect to the distance of the receiving node from the packet destination and its relative position to the sender. Hence, when a node broadcasts a packet it has to include in the routing header its distance from the destination, so that the nodes that receive it, can calculate their forwarding probability for it. Obviously, out of the neighbors of the sender node, that is to say, the nodes that are within range of its broadcast, the neighbor that is closest to the destination should be assigned a probability equal to one, to ensure the progress of the packet towards the destination. In what follows we interchange the terms "routing cost metric" and distance.

a. a linear forwarding probability function: Initially, we consider a linear decreasing function with probability values from 1 to 0. It is simple to observe that such a function satisfies the above requirements. This forwarding probability p is expressed by:

$$p = \frac{1}{d_{\min} - d_{\max}}\left[d\left(p_{\max} - p_{\min}\right) + p_{\min}d_{\min} - p_{\max}d_{\max}\right]. \tag{1}$$

where d is the distance between the candidate forwarder and the destination, $p_{\max}$ is the forwarding probability associated to the nearest possible candidate (i.e., for $d = d_{\min}$), and $p_{\min}$ is the probability associated to the furthest possible candidate (i.e., for $d = d_{\max}$). It is straightforward to observe that $0 \leq p \leq 1$ for $d_{\min} \leq d \leq d_{\max}$.

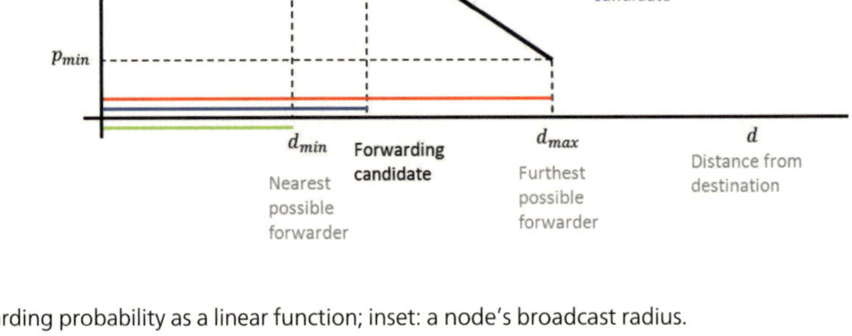

Figure 1. Forwarding probability as a linear function; inset: a node's broadcast radius.

http://dx.doi.org/10.5772/53997

This function provides differentiation between the forwarding probabilities of different nodes that receive the same packet in the broadcast range of the transmitter. However, it assigns a probability equal to 1 to exactly one node in each broadcast area, the node with d_{min} distance from the destination. Therefore, the packet's progress would heavily rely on that particular node. Moreover, if no node is to be found with this particular distance value in a topology, then there would be no certain forwarder for that packet in this broadcast area.

b. Piece-wise linear forwarding probability function: To deal with the case of no certain forwarders, the previous forwarding probability function can be modified to increase the number of potential forwarding candidates that can have high forwarding probability or equal to one. This can be achieved by using a piecewise function composed of an initial flat region where probability is one, followed by a decreasing linear function. The shape of the function is demonstrated in Figure 2. In this case, the forwarding probability is given by:

$$p = \min\left\{1, \frac{1}{d_{\min} - d_{\max}}\left[d\left(p_{\max} - p_{\min}\right) + p_{\min}d_{\min} - p_{\max}d_{\max}\right]\right\}. \quad (2)$$

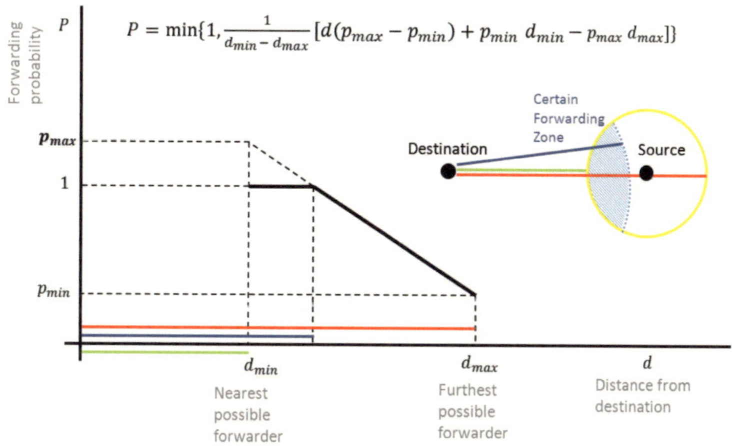

Figure 2. Piece-wise forwarding probability function and a node's broadcast radius.

This piecewise function is produced by setting $P_{max} > 1$ and hence introducing more than one certain forwarders. To guarantee the packets' progress to the destination, at least one neighbor with forwarding probability equal to 1 is needed and this is ensured by the flat region. Ideally this would be the neighbor on the shortest path to destination.

b. Binary forwarding probability function: The final variant we examine for the forwarding probability function is a 0-1 function. Figure 3 illustrates the step-wise function. In this case, nodes are either assigned a forwarding probability equal to one or a probability equal to zero, that is they either forward the packet always or they never do. Forwarding is again based on the value of the metric and there is a threshold value of the metric, which when surpassed it is decided that the node should not forward. Yet again, the number of forwarders may be increased or decreased by adapting the threshold value of the metric. We describe this function by:

$$p = \begin{cases} 0, & d_{threshold} < d \le d_{\max}, \\ 1, & d_{\min} \le d \le d_{threshold}. \end{cases} \quad (3)$$

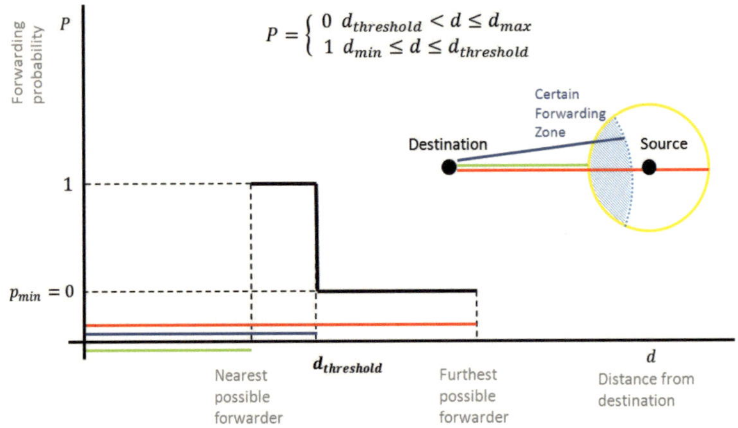

Figure 3. Step-wise forwarding probability function and a node's broadcast radius.

2.4. Back-off window differentiation

Having calculated its probability to forward a certain packet that it received, a node should proceed to decide when to do so. To this end, a randomized back-off mechanism is used, where each node calculates a window of back-off time slots and randomly selects a number of slots from that range. After this back-off timer has expired, the node will proceed to forward the packet with its predetermined probability. The focus here is to differentiate the back-off times of different nodes, focusing more on the ones with high forwarding probability. These highly-likely forwarders contribute significantly to the packet's progress towards the destination; therefore it is important to avoid collisions between their transmissions, making the packet move closer to the destination as fast as possible. To this end, back-off values are taken to be inversely proportional to a node's forwarding probability. However, the piece-wise and the step forwarding probability functions both include flat regions which can assign probability equal to 1 to more than one nodes which would result in them calculating the same back-off timers. For this purpose, the linear probability function (2) is used to calculate a base probability for each node which will then be used in order to calculate its back-off window *win*.

$$win = \left(win_{\max} - win_{\min}\right)\left(1 - p_{base}\right) + win_{\min}. \tag{4}$$

Where p_{base} is the base probability for that node and $win_{\min}$, $win_{\max}$ are values of minimum and maximum back-off that can be assigned, respectively.

2.5. Passive hob-by-hop acknowledgment and retransmission scheme

In order to acknowledge successful packet reception the forwarding procedure takes advantage of the broadcast nature of the wireless medium. After broadcasting a packet, a node can learn if at least one of its neighboring nodes received it by overhearing its neighbors' transmissions for a sort amount of time, thus also avoiding collisions. If a transmission of the last sent packet is overheard, then a node will drop the packet from its queue and continue to transmit the next. In case time goes by without overhearing any transmission of the last sent packet, then the node will retransmit it, as long as a maximum number of retransmissions has not been reached.

2.6. Multiple packets handling

Upon successful reception of a new non-expired packet, the node will have to calculate its forwarding probability for it, its back-off window for it and select a random back-off value from the range of the latter. A list with the packet and flow IDs of previously successfully forwarded packets can be kept to ensure that a node will not forward the same packet of a flow twice thus reducing redundant transmissions. After determining all of the above, a node will have to store individual packets according to the back-off timer that it has calculated for each of them and try to transmit them in time.

The manner in which the node will handle the various packets it has stored can be described as a system of multiple queues, each one containing packets for which the node has selected the same back-off value and each queue is a FIFO. After having successfully transmitted a packet and overheard its retransmission by a neighboring node, the node will look for the queue with the smallest back-off and pick the first packet from that. This ensures that a packet for which the node has a small back-off will have priority over one for which the node has a large back-off value. Taking into account that the back-off assignment favors the optimal forwarding candidates for a packet, this queuing policy allows the node to give priority to packets to whose progress it can contribute more.

2.7. Routing scheme adjustable parameters

a. Maximum probability ($p_{\max}$): The forwarding probability function slope defines the difference in the probability to forward the packet between the neighbors of the node currently holding the packet. Specifically, the steeper that slope is the more the neighbors closer to the destination will be favored. By setting the maximum probability to a value higher than 1, the slope of the forwarding probability function can be tilted, thus increasing the flat segment, which leads to more "certain" forwarders. This feature provides the forwarding function the ability to adapt in situations where more forwarders with high probability are needed.

b. Acknowledgment delay: After a node broadcasts a packet, it will start overhearing its neighbors' transmission in order to verify that the packet it sent has been broadcasted by one of them. The amount of time it can wait in this overhearing mode without success, until it decides it has to retransmit the packet, is called acknowledgment delay. If this interval is too small, then the node might end up retransmitting a packet that is successfully received by the further hops, thus adding one redundant transmission. On the other hand, if it is too long and no transmission is overheard, then the packet's progress will be delayed.

c. Time to live (TTL): To ensure that the packets will not circulate in the network long after they have reached the destination, a mechanism that renders them obsolete is needed. For this purpose, each packet has a fixed number of "credits" which are spent each time it is broadcast. These credits can be time units or number of hops traversed, under the assumption that a time unit equals the time it takes for the packet to move one hope further. A node that receives a packet with an expired TTL will discard it without calculating any forwarding probability or back-off window for it.

3. Simulation framework assumptions and setup

To model the behavior of our proposed opportunistic routing scheme and gain insight in its parameters in order to be able to optimize them, we developed a MATLAB-based platform as a time-driven simulator. The model we consider is time-slotted one. Specifically, a time-slot refers to the duration a fixed size packet needs to be broadcast and received. To setup the multi-hop wireless network simulation environment, we made a series of assumptions, regarding system information availability, packet propagation and channel errors.

For simulation simplicity we assumed a protocol propagation model i.e. a transmission can only be received, under some probability, by all nodes within a broadcast radius from the source. The nodes that lie within a node's broadcast radius are referred to as neighbors of that node. In our initial implementation all nodes in the network shared the same packet reception probability and same broadcast radius. Individualizing them is a straightforward programming exercise.

The simulation scenarios we include in the next section take place in a grid topology, such that a node may have four, eight or twelve neighbors, depending on the transmission radius (Figure 4). The nodes have fixed positions, known a priori, as far as calculating metric values is concerned. However, more randomized topologies are be examined as well, by setting random nodes of the uniform grid topology as inactive. From an implementation perspective, this was performed by randomly selecting nodes other than the source or destination of the packets and fixing their packet reception probability to zero.

http://dx.doi.org/10.5772/53997

For the purpose of testing the performance of the suggested procedure in realistic conditions, the assumption that there is full and accurate knowledge of nodes' locations in the topology was relaxed. Regardless of the metric used, the possibility of erroneous estimations of the metric value should be taken into consideration when testing the performance of a routing scheme, in order to examine its robustness.

To this end, a noise parameter in the metric calculation was introduced in the simulation environment. We consider that the calculated metric is taken randomly in an interval centered on the real metric value and spans by a percentile which is a simulation parameter.

Experiments were conducted on a 40 x 40 node grid topology to measure delay, loss ratio and resource consumption for varying network densities and channel error conditions. Each experiment with a given set of parameters was repeated for 100 runs and the results presented in what follows are averages over the number of runs. Delay measurements were performed on the shortest path from source to destination and a classic single-path lowest-cost route has been simulated and used as the basis of comparisons.

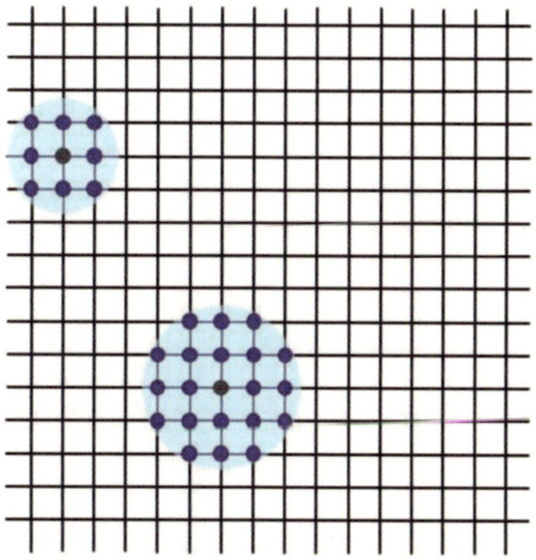

Figure 4. Grid topology with transmission radii for 8 and 12 neighbors.

4. Results

4.1. Tuning the schemes' adjustable parameters

4.1.1. The effect of the forwarding probability function

The forwarding probability function slope defines how great the difference in the probability to forward the packet will be between the neighbors of the node currently holding the packet. Specifically, the steeper that slope is the more the neighbors closer to the destination will be favored. To ensure the packet's progressing to the destination, at least one neighbor with forwarding probability equal to 1 is needed; ideally this would be the neighbor on the shortest path to destination. This leads to a linear forwarding probability function. As the packet error ratio (PER) increases, more "certain" forwarders are needed, to make up for failed packet receptions, but the need arises to relax potential collisions among them.

By setting the maximum probability to a value higher than 1, we tilt the slope of the forwarding probability function, thus increasing the number of certain forwarders. Figure 5 illustrates how tilting the slope increases the number of certain forwarders in an 8 neighbor topology. Initially there is only one certain forwarder, the one that is closest to the destination. Two more forwarders are added that are closer to the destination than the node that transmitted the packet. If the slope is increased, two more forwarders are added, that are slightly further away from the destination, which will make the packet progress diverge sideways from the shortest path. Values of maximum probability in [1.4, 2] contribute to the packet's progress without diverging much from the shortest path. Furthermore, for a given PER value, having more than one certain forwarders yields lower delay.

This is verified by Figure 6 which illustrates how tilting the forwarding probability function's slope leads to lower delays, in the context of a 10-hop shortest path. It should be noted that only the maximum forwarding probability parameter is examined at this point; lower delay can be achieved by adjusting the maximum back-off window value as well, which in this case is variable, dependent on the forwarding probability, taking values in [1, 8]. Nonetheless, our probabilistic scheme outperforms single-path routing for PER values higher than 0.25.

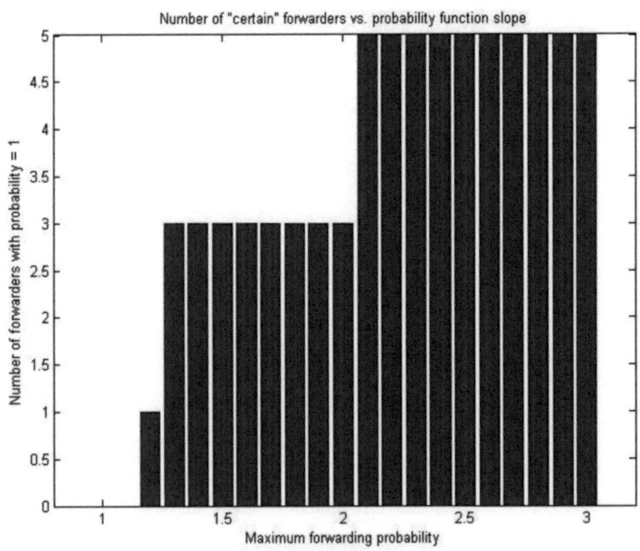

Figure 5. Increasing the number of certain forwarding nodes tilting the probability function slope.

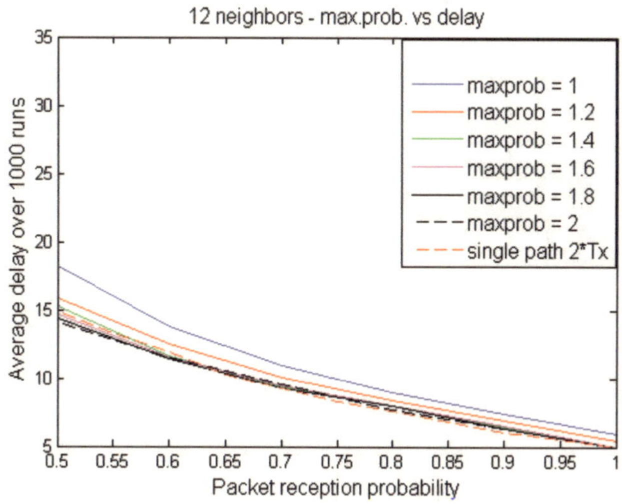

Figure 6. Tilting the slope to create a saturated, piece-wise probability function yields reduced delays.

4.1.2. Back-off window differentiation

There were two general approaches to back-off window schemes with respect to the window range values. Firstly, in the fixed back-off window option, all transmitting nodes would randomly select their back-off values from the same range of window values. Secondly, when differentiating, each transmitting node randomly selects its back-off value from a different range of numbers. Specifically, the back-off window of a node will be randomly chosen between a smaller set of numbers, the larger its forwarding probability is, so as to reduce delays. Figure 7 below illustrates the delay performance of the fixed (7a) and differentiated back-off window schemes (7b), for increasing packet error rate values, in a scenario where the source is 10 hops away from the destination. There is a steeper increase in delay for the fixed back-off window scheme as the width of the back-off window interval increases, which renders fixed back-off values larger than 2 inefficient.

It can be observed that the lowest delay is measured for a back-off window of 1, which raises the question, why differentiate between nodes at all. The reason is that back-off differentiation also yields lower resource consumption.

To capture the effect a packet's transmission has on the network, we track the footprint its transmissions produce over time on the nodes as it is forwarded towards the destination, until all transmissions cease. The times each node has received the packet are averaged over the number of the different runs of the experiment.

For all results referring to footprints hereon the source node's coordinates are (10,20) and the destination is at (20,20). Figure 8 illustrates the footprint for a flow with back-off window equal to 1. As we increase the width of the back-off window the flooding is limited to an area around the shortest path. This is shown by the plots in Figure 9. It should be noted that a linear forwarding probability function with maximum probability equal to 2 was used for these experiments. There is a trade-off between low delay performance and resource consumption which should be addressed by having each flow's specific requirements in mind. For example, in a network where only one flow is present at a time, a back-off window set to 1 would yield the lowest delay possible, whereas in the presence of multiple flows, a more conservative back-off scheme with the window interval set to [1,8] should be used.

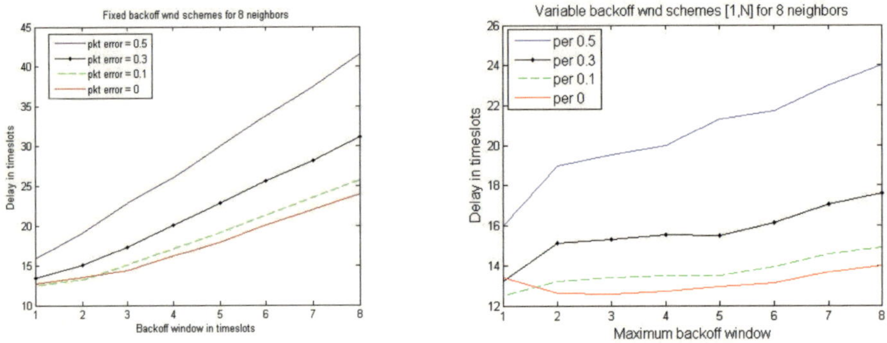

(a) (b)

Figure 7. Delay in number of slots for a 10-hop shortest-path source-destination pair for the fixed (a) and variable (b) back-off schemes.

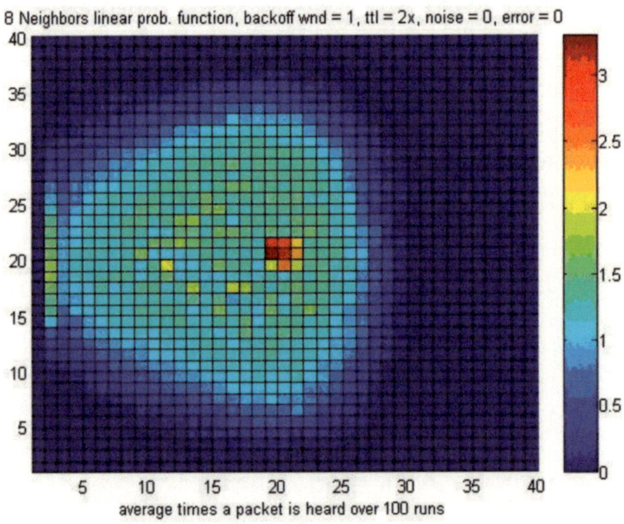

Figure 8. A back-off window of one time slot yields low delay compared to a variable backoff window but has a large footprint i.e. is resource inefficient.

http://dx.doi.org/10.5772/53997

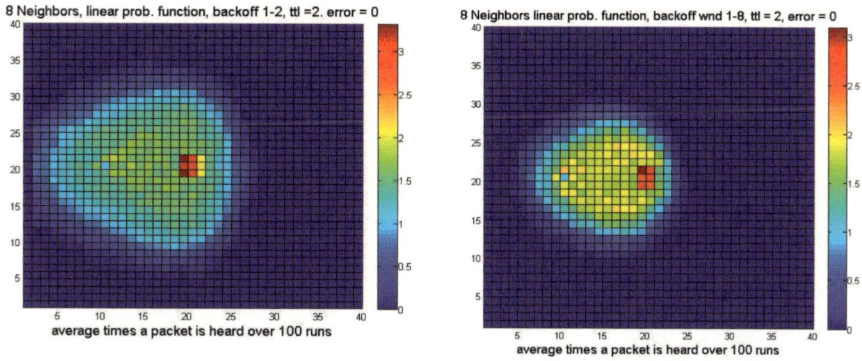

(a) (b)

Figure 9. Increasing the back-off window range from [1,2] in (a) to [1,8] in (b) under a piece-wise probability function reduces the footprint causing less interference.

4.1.3. Metric miscalculation error

To evaluate the robustness of our scheme to mistakes in metric estimations, we examine the effect of introducing metric noise.

1. Linear probability function: To measure resource consumption for the linear probability function with maximum probability set to 2, we track the flow's course in a scenario where there is no channel error and the source is 10 hops away from the destination and use a balanced back-off window scheme in the interval [1, 4].

Figure 10 below illustrates how the linear forwarding probability scheme reacts to increasing metric miscalculation. It is noteworthy that the linear forwarding probability scheme is insensitive to metric noise up to 10% of the accurate metric's value and performs decently even at the presence of noise equal to 30% of the accurate metric's value. Having in mind that the metric values are randomly chosen from a uniform distribution in such wide intervals, it is obvious that metric noise equal to 0.3 already presents a scenario of extremely inaccurate metric estimation.

2. Step-wise probability function: We compare the performance of the linear probability function in the presence of noise to that of the step-wise probability function. For the purposes of these experiments the step was set to 0, such as the neighbors that have a smaller distance from destination, than the node currently holding the packet, will forward the packet. The source is again at (10,20) and the destination at (20,20) and the back-off scheme is again set to [1, 4] interval.

Figure 11 illustrates the step-wise function's performance under increasing metric noise conditions. For the same values of metric noise, the step-wise function consumes fewer resources than the linear function and it is also less affected by metric noise. This is justified by the fact that even when inaccurate estimations are made, only a few nodes will calculate their metric so that they fall below the threshold, whereas the majority of nodes will get the same probability as what they would get if there was no noise.

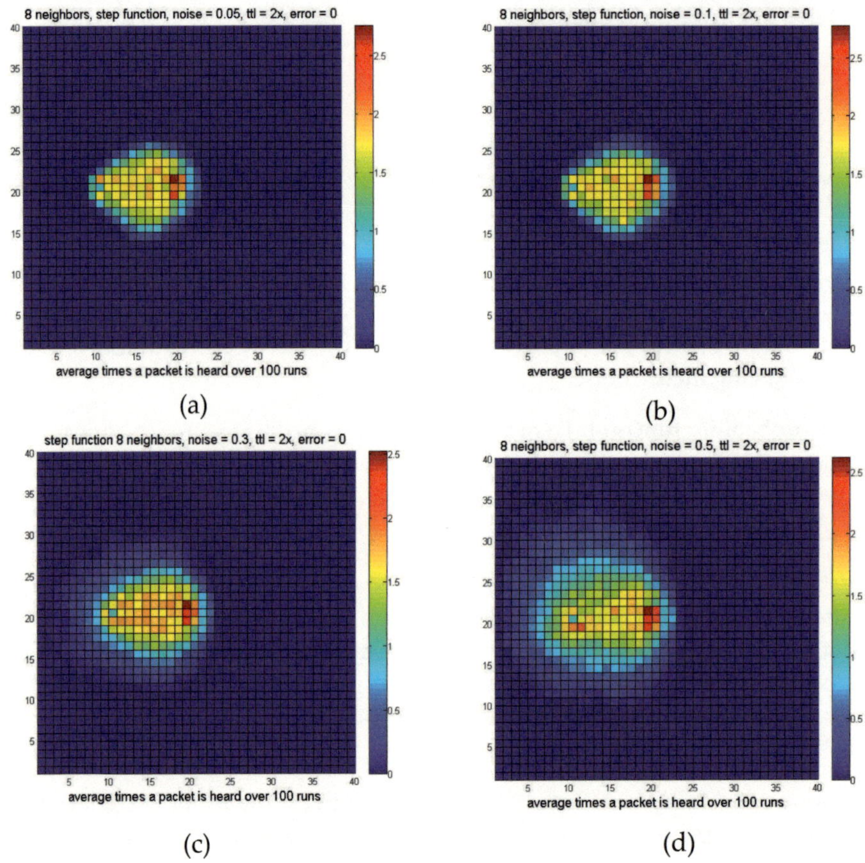

Figure 10. Increasing the metric miscalculation factor from 0.05 (a), 0.1 (b), 0.3 (c), to 0.5 (d) slightly affects the resource efficiency for an 8-neighbor topology under a piece-wise probability function.

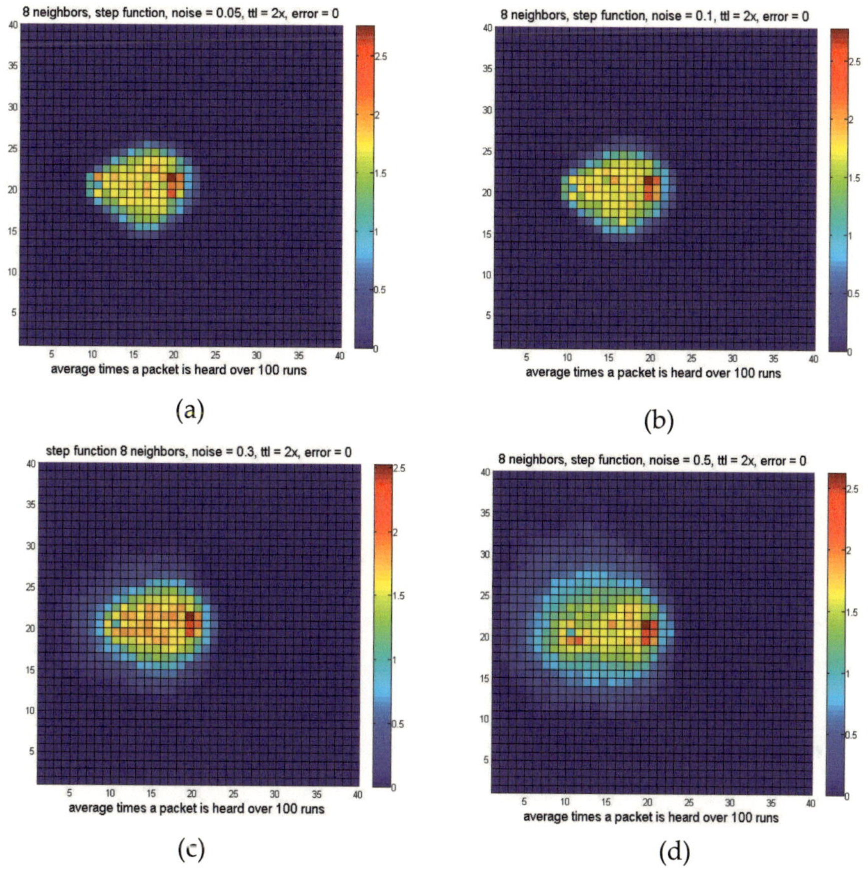

Figure 11. Increasing the metric miscalculation factor from 0.05 (a), 0.1 (b), 0.3 (c), to 0.5 (d) slightly affects the resource efficiency less than in figure 10 when using a step-wise probability function.

4.1.4. Multiple flow support

It is of interest to indicate that the routing scheme we have devised thus far is able to support the interaction of different flows in the network and examine their behavior. The scenarios in the following figures were chosen with respect to the most common cases presented in wireless mesh and sensor networks. Figure 12 illustrates the scenario of two nodes sending packets to the same destination node, whereas Figure 13 is the reverse, which can be considered as a "downlink" case. Both provide evidence that the system supports multiple flows and that the scheme behaves as expected both in terms of packet error, as well as in terms of back-off delay.

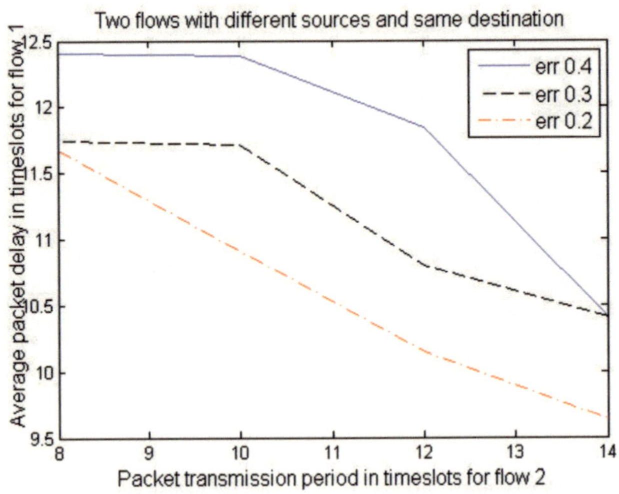

Figure 12. Delay performance of a 50 packet flow while competing with another for the same destination.

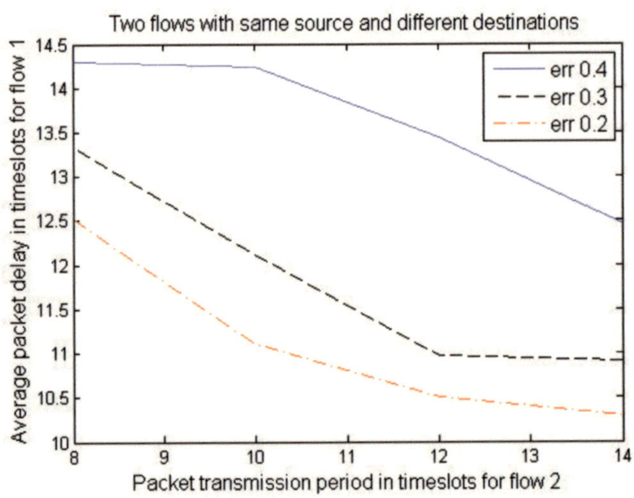

Figure 13. Delay performance of a 50 packet flow while competing with another flow from the same source.

4.2. Performance comparison

4.2.1. Comparing against SOAR

For our comparative simulations, SOAR's algorithm for packet forwarding decisions was used, combined with the proposed passive acknowledgment scheme, in order to test the performance of its opportunistic features. SOAR initially uses ETX as a metric in order to decide on the cost of forwarding, however, for comparison purposes, hop distance was used for both protocols. SOAR behaves similarly to shortest path routing in no-error conditions, constraining the flow along the shortest path from the source to the destination, as shown in Figure 14. When metric miscalculation is present, SOAR's delay increases signifantly, as opposed to the proposed scheme's performance which is unaffected, as shown in Figure 15. This can be explained by the quasi-deterministic forwarding scheme used by SOAR. If metric miscalculation occurs at the source (who creates the list of forwarders), the error will propagate along with the list, since it is included in any sent packets. Therefore consequent calculations based on this list will be influenced by even a single miscalculation error.

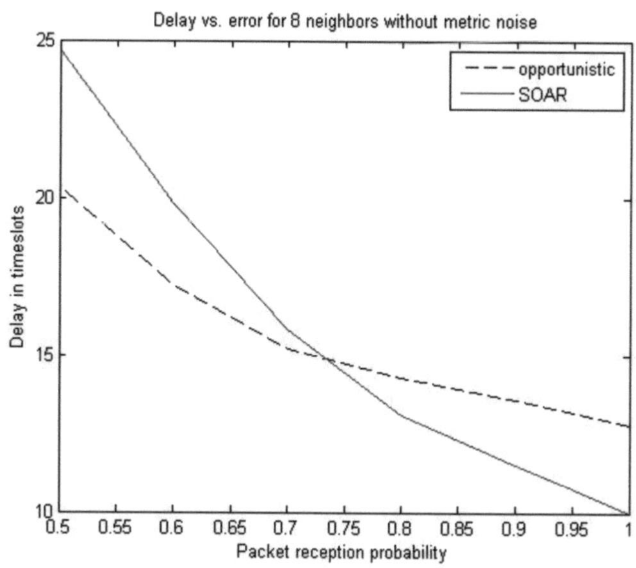

Figure 14. The proposed scheme (dashed) yields lower delays than SOAR (solid) for packet error probability over 0.25.

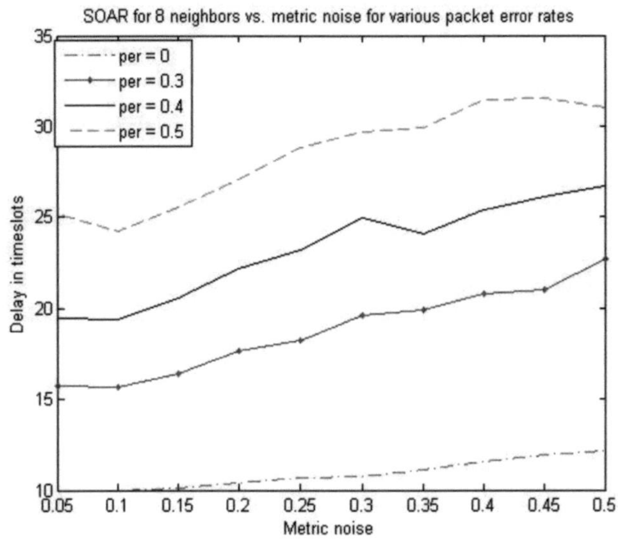

Figure 15. SOAR's delay performance degrades as metric miscalculation is increased.

4.2.2. Comparing against direct transmission

Directed Transmission is a parametric probabilistic routing protocol focuses on design simplicity, distributed routing decisions and robustness to metric miscalculation. It should be noted that Directed Transmission does not account for losses due to poor link quality, so the two protocols were compared in no-channel error scenarios, where metric miscalculation was present. When error in metric calculation increases, directed transmission's use of resources does not increase significantly, as is the case with the scheme proposed. However, the spread of the flow on the grid is comparable to its equivalent in the proposed scheme when the piece-wise probability function was used and larger than its equivalent when the step-wise probability function was used. Furthermore directed transmission has a larger average number of transmissions needed to deliver a packet along the 10 hop path, regardless of metric noise. These are depicted in Figure 16. This demonstrates that a routing protocol can be simple enough as the one we proposed and at the same time conserve resources sufficiently to be applied in WSNs without suffering from low delay performance.

http://dx.doi.org/10.5772/53997

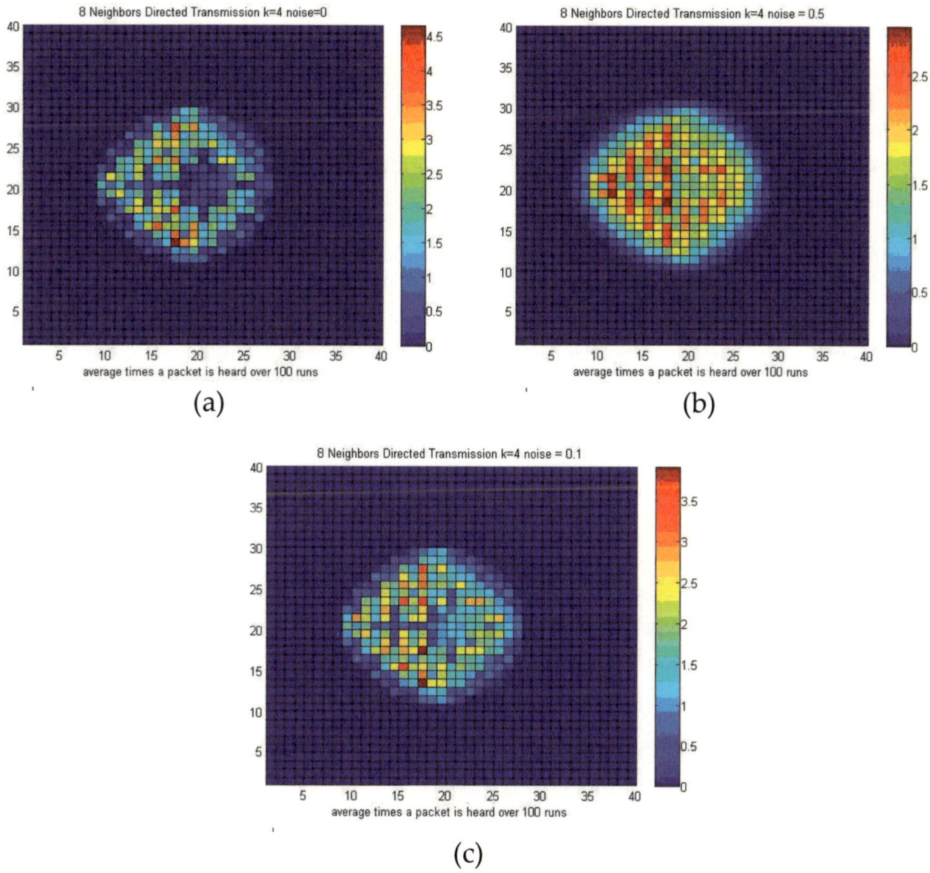

Figure 16. Directed Transmission appears equally robust to metric miscalculations as the proposed routing scheme, still compared to Figures 10 and 11 it requires on average more retransmissions to deliver a packet.

5. Conclusion

We have addressed the two key design decisions of designing an opportunistic routing scheme. How nodes should decide whether to forward or not and when is the most efficient time to do so. We gave indications using a simulation framework on how the forwarding decisions and transmission timing alone affect performance. This was done through a probabilistic forwarding scheme, whose parameters can be tuned to allow for low resource consumption and high delay performance, while being robust to misinformation. The overall routing protocol comprising the forwarding scheme along with the timing and the acknowledgement mechanism is stripped of complex routing mechanisms and so we examined which channel error conditions and topology density it is beneficial to use opportunistic routing instead of traditional routing.

Simulation results demonstrate that the suggested opportunistic scheme can outperform single path routing for error values larger than 15% -20%, for a well-selected slope of the forwarding probability function, with restrained use of resources. Furthermore, we have shown that the optimal manner of adapting to increasing error is to increase the number of forwarders by increasing the slope of the forwarding probability function. In particular, the number of certain forwarders has the most impact on performance and they need to be increased with respect to error conditions.

To reduce resource consumption, in terms of packet transmissions, utilizing a step-wise function is a sound approach, which proved robust to metric miscalculations. Finally, there is a tradeoff between differentiating each forwarder's back-off value to reduce resource consumption and reducing delay. Simulation results show that a variable back-off scheme that gives priority, by means of smaller back-off windows, to best forwarders according to their forwarding probability is preferable to a fixed back-off window for all forwarders.

Acknowledgements

Dr. Angelakis and Ms. Gazoni were with ICS-FORTH, Greece when conducting part of the research presented and acknowledge the aid of Dr. V. A. Siris and the support of the FP7 project EU-MESH.

We further acknowledge the support of the FP7-People-2007-3-1-IAPP-218309 project. This work has been also partially supported by CENIIT at Linköping University, the Swedish excellence center ELLIIT, and the Swedish Research Council (VR) through project B0581701.

We finally thank Mr. Lei Lei for his help in the final revision of the material.

Author details

Vangelis Angelakis[1], Niki Gazoni[2] and Di Yuan[1]

1 Department of Science and Technology, Linköping University, Norrköping, Sweden

2 Forthnet S.A., Systems Engineering and Design Department, Athens, Greece

References

[1] C. L. Barrett, S. J. Eidenbenz, L. Kroc, M. Marathe, and J. P. Smith. Parametric probabilistic sensor network routing. In proc. ACM WSNA '03, doi 10.1145/ 941350.941368, 2003.

[2] S. Biswas and R. Morris. Opportunistic routing in multi-hop wireless networks. In proc. SIGCOMM05, doi: 10.1145/972374.972387, 2005.

[3] J. Du, H. Liu, and P. Chen. Omr: An opportunistic multi-path reliable routing protocol in wireless sensor networks. In proc. IEEE ICPPW '07, doi: 10.1109/ICPPW.2007.61, 2007.

[4] C. E. Perkins and P. Bhagwat (1994). Highly dynamic destination-sequenced distance-vector routing (dsdv) for mobile computers: ACM SIGCOMM Computer Communication Review, doi: 10.1145/190809.190336.

[5] N. Gazoni, V. Angelakis, V. Siris, and R. Bruno. A framework for opportunistic routing in multi-hop wireless networks. In proc. ACM PE-WASUN, doi: 10.1145/1868589.1868600, 2010.

[6] S. Jain and S. Das. Exploiting Path Diversity in the Link Layer in Wireless Ad Hoc Networks, in proc. IEEE WoWMoM '05, doi: 10.1016/j.adhoc.2007.07.002, 2005.

[7] D. B. Johnson, D. A. Maltz, and J. Broch (2001). Dsr: The dynamic source routing protocol for multi-hop wireless ad hoc networks. In: Charles E. Perkins editor, Ad Hoc Networking, Addison-Wesley. pp. 139-172.

[8] De Couto D. S. J, Morris R., Aguayo D., and Bicket J. A high-throughput path metric for multi-hop wireless routing. In proc. ACM MobiCom '03, doi: 10.1145/938985.939000, 2003.

[9] Z. J. Haas, J. Y. Halpern, and L. Li (2006). Gossip-based ad hoc routing. IEEE/ACM Transactions on Networking (TON) doi:10.1109/TNET.2006.876186.

[10] L. Haitao, et al. (2009). Opportunistic routing for wireless ad hoc and sensor networks: Present and future directions. IEEE Communications Magazine, doi: 10.1109/MCOM.2009.5350376.

[11] S. Mueller, R. Tsang, and D. Ghosal. (2004) Multipath routing in mobile ad hoc networks: Issues and challenges. In M. C. Calzarossa, E. Gelenbe, editors, Performance Tools and Applications to Networked Systems, Springer-Verlag pp. 209-234.

[12] M. S. Nassr et al. Scalable and Reliable Sensor Network Routing: Performance Study from Field Deployment. In proc. IEEE INFOCOM '07, doi: 10.1109/INFCOM.2007.84, 2007

[13] R. Rajaraman, (2002) Topology control and routing in ad hoc networks: A survey. ACM SIGACT News, doi>10.1145/564585.564602.

[14] E. M. Royer and C.E. Perkins. An implementation study of the aodv routing protocol. In proc IEEE WCNC, doi: 10.1109/WCNC.2000.904764, 2000.

[15] E. Rozner, J. Seshadri, Y. Mehta, and L. Qiu. SOAR: Simple Opportunistic Adaptive Routing Protocol for Wireless Mesh Networks. In IEEE Trans. On Mobile Computing 8(12): 1622-1635, 2009.

[16] R.C. Shah, S. Wietholter, A. Wolisz, J. M. Rabaey. When does opportunistic routing make sense? In Proc. IEEE Percom, doi: 10.1109/PERCOMW.2005.90, 2005.

[17] J. Wu, M. Lu, and F. Li. Utility-Based Opportunistic Routing in Multi-Hop Wireless Networks," Proc. IEEE ICDCS '08, doi: 10.1109/ICDCS.2008.90.

[18] Y. Yuan, H. Yang, S.Wong, S. Lu, and W. Arbaugh. Romer: Resilient opportunistic mesh routing for wireless mesh networks. In proc IEEE WiMesh workshop 2005, SECON, pp. 146-158, 2005.

[19] K. Zeng, W. Lou, and H. Zhai. On end-to-end throughput of opportunistic routing in multirate and multihop wireless networks. In proc. IEEE INFOCOM '08, doi: 10.1109/INFOCOM.2008.133 2008.

[20] Z. Zhong and S. Nelakuditi. On the efficacy of opportunistic routing. In SECON '07: In proceedings of the 4th IEEE International Conference on Sensor and Ad Hoc Communications and Networks, doi: 10.1109/SAHCN.2007.4292856, 2007.

[21] S. Chachulski et al., Trading Structure for Randomness in Wireless Opportunistic Routing. In proc. ACM SIGCOMM '07, doi: 10.1145/1282427.1282400 2007.

[22] P. Jacquet, T. Clausen, (2003) Optimized Link State Routing Protocol (OLSR), RFC 3626

[23] Y. Yan et al. Practical Coding-Aware Opportunistic Routing Mechanism for Wireless Mesh Networks. In proc. IEEE ICC2008 doi: 10.1109/ICC.2008.541,.2008.

[24] Z. Zhong et al., (2006) On Selection of Candidates for Opportunistic AnyPath Forwarding. ACM SIGMOBILE Mobile Computing and Communications Review doi>10.1145/1215976.1215978.

Chapter 5

Reducing Routing Loops Under Link-State Routing in Wireless Mesh Networks

Takuya Yoshihiro and Masanori Kobayashi

Additional information is available at the end of the chapter

http://dx.doi.org/10.5772/53814

1. Introduction

In Mobile Ad Hoc Networks (MANETs), one important issue is how to provide stable communication between nodes against transition of network state such as node mobility or quality transition of communication links. In MANETs, due to the fragile nature of wireless links, routing protocols are designed to be more robust and resilient against failure, while restraining the network load of control messages even if nodes are distributed densely. Four routing protocols, i.e., AODV[1], DSR[2], OLSR[3], TBRPF[4] have been standardized so far. Although each of which has its own mechanism that is convenient for MANET, they do not still displayed a sufficient performance to be applied in practice.

One of the drawbacks in these routing protocols is that they do not consider the transition of link quality in their process of computing forwarding paths; they use hop counts as their basic criteria to compute forwarding paths of packets. However, the transition of link quality including link failure is not avoidable in MANET because its basic requirements include node mobility and wireless links. To take this important characteristic into account, many routing metrics have been proposed that quantifies link quality from various points of view [6]-[11]. For MANETs in which mobility is included in general, the event that affects the most on communication performance is link failure due to mobility. Thus, most of the proposed link metrics tries to quantify the probability of link failure by way of measuring node speed, RSSI (Received Signal Strength Indication), and so on [6]-[11]. For Wireless Mesh Networks (WMN) [5] in which nodes are stationary, because the risk of link failure is far smaller than MANET, the link quality that should be quantified is the quality of communications such as communication speed, delay, and stability of links [12]-[16]. For example, ETX (Expected Transmission Count) [12], which is one of the most commonly used link metrics, quantifies the average transmission count in 802.11 MAC computed from success ratio of MAC transmission, and ETT (Expected Transmission Time) [13] extends ETX to quantify the average transmission time of a MAC frame in the link.

As far as proactive link-state routing such as OLSR is concerned, it is well understood that introducing dynamic link metrics make networks far robust and resilient, and consequently improve performance of networks in practical situations. However, simultaneously, such dynamic metrics cause communication paths to be changed frequently. Note that the paths flapping behavior is not always bad, because it is the result of continuous effort of routing protocols to find better quality paths. Nevertheless, it certainly increases the risk of several inconvenient phenomena such as packet looping.

Packet looping is one of the very harmful problems because looping packets travel along the same link repeatedly and consume significant capacity of the network. In general, larger number of looping packets appears when the network topology changes including link metrics more frequently. In other words, dynamic metrics by nature involves the risk of this kind of instability in exchange for the flexibility against wireless instability. Therefore, it is one of the goals for us to reduce the harmful influence of packet loops, while simultaneously holding the flexibility brought from dynamic metrics.

Note that, in wireless multi-hop routing, there are several causes of reducing communication performance other than packet looping, and they are deeply related with one another. Not only packet loops, but also congestions due to interference, and further link failures due to wireless instability or mobility are also regarded as the essential elements that should be considered in MANET routing schemes. Especially, interference would be the most focused element in the current state of the art. However, in wireless networks, packet looping and interference are deeply related with each other so that improving performance from the viewpoint of looping would also be an important part of the contribution.

In this paper, we present a new loop reduction method for Wireless Mesh Networks, which follows the description of the literature on loop-free techniques for wired and wireless networks. This article is organized as follows. In Section 2, we first describe a concise description of proactive link-state routing protocols and related techniques in WMNs, including dynamic metrics. In Section 3, we review the literature of loop prevention methods for wired networks. In Section 4 we describe the loop prevention methods proposed for WMNs. Then, in Section 5, we present a new loop prevention method and its evaluation results. Finally in Section 6 we conclude the article.

2. Packet looping problem and its harmful influences

Packet looping is a harmful phenomenon in which packets are forwarded among the same nodes. Looping packets significantly consume resources of networks, and consequently cause severe congestion. Packet looping traditionally has been discussed in wired networks, where looping occurs typically when a link fails. Link failure triggers the process of paths re-computation in routing protocols. Then, in the transient state to converge to the new shortest paths state, routing tables among some nodes possibly create loops temporarily. This kind of looping is traditionally called as "routing-table loops" or just "routing loops."

See Figure 1 for an example of a routing loop as a result of link failure. Figure 1(a) shows an initial link metrics on the circular network with three nodes. The shortest paths from *A* (resp. *B*) to *C* computed by *A* itself (resp. *B* itself) is also indicated here. Suppose that link *(A, C)* fails. With this topology change, the shortest paths are finally changed to the state shown in Figure 1(c). However, in the transient state where only *A* knows the link failure and *B* does not know it (Figure 1(b)), A forwards packets destined to *C* using the next hop *B*, while *B* forwards those using the next hop *A*, then they loops between *A* and *B*. This loop continues until *B* updates its next hop nodes and consumes significant network resources of *A* and *B*.

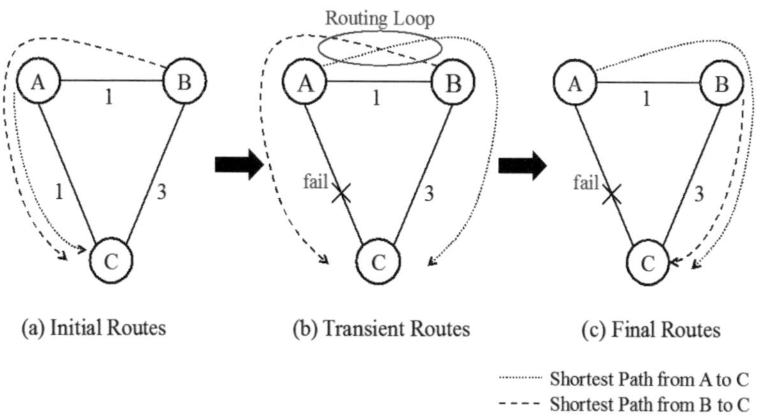

Figure 1. A Routing Loop in Case of Link Failure

Routing loops caused of failure occurs in both wired and wireless networks. However, wireless network has another type of risk for routing loops. In wireless multi-hop networks with proactive link-state routing schemes such as OLSR, it is general to deploy a dynamic metric to improve the performance of networks over instable wireless links. With dynamic metrics, frequent changes of link metrics arise to be a major cause of routing loops.

The typical example of loops caused from metric change is shown in Figure 2. In Figure 2(a), the initial metrics are shown with the same topology as in Figure 1. The shortest paths from *A* and *B* to *C* is the direct path to *C*. Suppose that metrics of links *(A, C)* and *(B, C)* are simultaneously raised to 3. In the final state shown in Figure 2(c), the shortest paths are the same as the initial state. However, in the transient state shown in Figure 2(b) where *A* knows the metric change of only *(A, C)* and *B* knows the metric change of only *(B, C)*, the shortest path for *C* computed in *A* is via *B* while the one computed in *B* is via *A*, then the packets loop between *A* and *B*. You may imagine that the situation shown in this example may be rare in practice. However, similar situation frequently occurs under dynamic metrics. Imagine the case heavy traffic suddenly appears in a network. Dynamic metrics sensitively respond to the traffic load and then many metric values would be raised simultaneously.

Note that the routing loops cause severe degradation of network performance. Speakman et al. [17] measured the impact of packet looping through their proposed method that detects looping packets at each node and drops them immediately. From the evaluation with mobility scenarios, they found the throughput of the network is improved by at most 10%, which shows the significant occupancy of resources with looping packets.

In wireless multi-hop routing, there are several causes of reducing performance other than packet looping, and they are deeply related with one another. Generally, the cause that has the largest impact on the performance would be the interference among radios. Because in wireless networks radio spreads for all directions, packet transmission on a link can be disturbed by other link's transmissions. Especially, in case of multi-hop networks over 802.11 MAC, hidden terminals significantly degrade the network performance by generating severe congestion. If such congestion is extreme, naturally wireless links are to fail. Link failure in turn invokes paths flapping, and the path flapping creates loops. Then, the loops again grow up congestions. In this way, the circulation of those harmful influences is formed. To reduce these harmful influences, it is essential to take measures for each of the causes. As one of the measures to deal with this situation, the techniques to reduce routing loops would be an important part to improve the performance of wireless multi-hop networks.

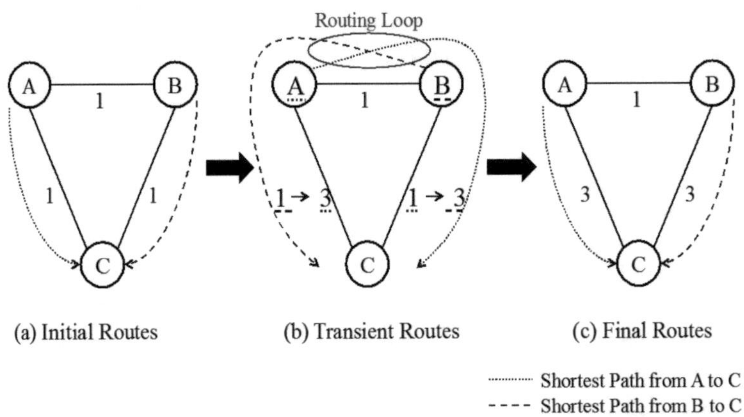

Figure 2. A Routing Loop in Case of Metric Changes

3. Literature on loop-prevention techniques for wired networks

3.1. Loop-prevention in distance-vector schemes

We start with the literature of loop prevention techniques for wired networks. To construct loop-free paths for every destination of a given network is a primary routing problem in the Internet. The most general approach for this problem is to use shortest-paths as the forwarding path for each destination. There are two major routing strategies, e.g., Distance-Vector schemes and Link-State schemes, which have been deployed in the Internet for a long time.

Distance-Vector routing schemes are deployed from the early stage of the Internet, and one of them is standardized as the representative routing protocol called RIP (Routing Information Protocol) [19]. RIP is based on the distributed Bellman-Ford algorithm [18]. Distance-vector routing scheme is driven with a simple mechanism: each node maintains a distance-table in which the distance for each destination is held, and advertises the distance-table to every neighbor periodically. Each node has only to choose the neighbor that has the shortest distance for each destination to construct its routing table. Unfortunately, this simple scheme has a serious problem so called the count-to-infinity problem [18]. The count-to-infinity problem occurs in case of topology change such as link failure, where the distance for a destination increases repeatedly among involved nodes until reaching the maximum distance defined in the protocol. In this period of time, packets are forwarded among those involved nodes and loop among them.

Several solutions are presented to reduce the harmful influence of the count-to-infinity problem. As a simple solution deployed in the early days, techniques so called split-horizon and poison-reverse [18] are well known. Although they can effectively reduce the affect of count-to-infinity phenomenon, the influence of the problem is not still negligible. One of the early-days solutions for this problem is to exchange full paths information in the routing scheme. This approach is currently known as Path-Vector routing, which is still used in inter-AS routing protocol BGP (Border Gateway Protocol) [21]. (Note that BGP is currently deployed not only for preventing loops, but also for several other functionalities such as controlling policies.) However, maintaining full paths information for every destination requires so much cost. To reduce the cost of control messages, Cheng et al. proposed a method to limit the information of forwarding paths to destinations that is transmitted into the network, with which they still prevent count-to-infinity problems [22]. Unfortunately, although this method mostly prevents routing loops, it cannot still eliminate temporary routing loops completely.

The first solution that prevents count-to-infinity problem without using path information was proposed by Garcia-luna-aceves, which was called as DUAL [23]. In [23], he presented locally computable sufficient conditions to be loop-free when a node changes its successor (next-hop) nodes to forward packets. Namely, when a node wants to change its successor node for a destination, it firstly checks the condition of loop-freedom. If the condition is met, it safely changes the successor node. Otherwise, it invokes a diffusing computation by sending a query to its neighbors to find a feasible successor. A neighbor node that received the query again send a query to its neighbors if it does not have a feasible successor, and when it received all the responses from its neighbors, it surely finds a feasible successor and returns the response with the successor information to the sender of the query. When the sender received the responses from all neighbors, it determines the new feasible successor. This coordination of nodes guarantees the selection of a "safe" successor to be loop-freedom at any instant. The mechanism of DUAL is implemented in the routing protocol EIGRP (Enhanced Interior Gateway Protocol) [24].

DUAL is improved in [25], in which only one-hop query processing is required, by means of using the predecessor information (i.e., information of forwarding paths to reach destinations) in the similar way as BGP. Later, Schmid et al. proposed to prevent the count-to-infinity problem without changing the message format of RIP routing protocols, if all link costs in a network is uniform [26].

3.2. Loop-prevention in link-state schemes

Link-state routing is another representative routing strategy that is also deployed from the early stage of the Internet. In this category of routing family, IS-IS (Intermediate-Systems, Intermediate-Systems) [27] and OSPF (Open Shortest Path Fast) [28] are the representative standardized routing protocols. In the link-state routing schemes, every node advertises the neighbor information (i.e., link information) to have all nodes in the network share the same image of whole network topology, and then every node computes the shortest paths on the shared network topology. Because all neighbor information is advertised through the network, link-state routing schemes require larger cost of control messages than distance-vector routing schemes.

Link-state routing schemes exchange more information among nodes, and it significantly reduces the time for path converging. Nevertheless, the risk of routing loops still remains in the face of topology changes. DUAL [23], which we described as a loop-free technique for distance-vector algorithms, again can be applied for link-state routing schemes so that link state routing works without routing loops under the diffusing computation. However, because this mechanism is based on distance-vector schemes and suitable to use on it, implementing it over link-state scheme is a little complicated. So, as a simple and feasible method to perform loop-free convergence to the new state of routing tables against single link/node/SRLG (Shared Risk Link Group) failure, Francois et al. proposed a method to control the order of updating routing tables not to create loops without any additional messages [29]. In their method, when a router computed a new successor to forward packets, it waits for a while until neighbor routers no longer select it as their successors, before updating its successor. To shrink the waiting time, simple messages to guarantee the order of updating successors can be used. Francois et al. further proposed a method to perform a planed link failure without loops by increasing a link cost gradually until no flow uses the link [30]. The mathematical analysis of the same problem is seen in [31]. In this method, they present algorithms to compute the sequence of intermediate values of link costs with which we can change a link cost to an arbitrary value without loops at any instant. With this sequence of link costs, it is possible to control the amount of traffic on the link, i.e., we can decrease or increase the traffic on a link gradually without rapid change of the link load.

4. Literature on loop-prevention techniques for wireless networks

4.1. Loop-prevention in reactive routing schemes

For wireless networks, generally two routing strategies are deployed, i.e., proactive and reactive routing. Proactive strategy takes a similar mechanism as wired networks; it always maintains a routing table that includes next-hop nodes for all destinations. In Contrast, reactive strategy is a new approach for MANET in which network load of control messages are reduced by searching forwarding paths to create its routing table entries on demand.

Reactive routing schemes such as AODV searches a path when a request for data delivery occurs, and by that they save the network load of control messages for unnecessary destinations. Thus, the family of reactive routing schemes is suitable for the case where communication amount or destinations to deliver data are limited. Reactive routing schemes are by nature loop-free because, once the path is determined, it is basically used unless a link or a node on the path fails. However, they require a path repairing process when the path is broken, which inevitably degrades the performance. To overcome this degradation, the method called ROAM [32] is proposed that allows nodes to change their successors without any message exchange. By means of using loop-free conditions given in [23], ROAM also guarantees loop-freedom with the more flexible paths control functionality. Also, [33] presents a method to improve the efficiency of path repairing in reactive routing schemes using the loop-free condition given in [23].

On the other hand, proactive routing schemes such as OLSR always maintain a path for every destination using routing tables. Note that the majority of the proactive routing schemes proposed so far are based on link-state strategy. In the link-state routing, every node in a network advertises its neighbor information (i.e., link information). As a result, all nodes in the network share the topology of the network, from which nodes compute their routing tables. Proactive routing schemes are able to begin communication without delay of searching paths, whereas they requires constant load of control messages. Because the load of control messages in dense networks is significant, OLSR deploys an effective load reduction technique called MPR (Multi-Point Relay), which limits the relay nodes in the flooding procedure that advertises messages throughout a network.

4.2. Dynamic metrics for proactive networks

In proactive link-state routing schemes, routing tables have to be maintained so that better paths are always available over the transition of wireless link quality. However, the initially standardized routing protocols including OLSR do not take it into account because they compute the shortest paths with respect to hop-count. To achieve more flexible routing over unstable wireless links, dynamic metrics were introduced into the shortest-path routing.

For wireless mesh networks, De Couto, et al. first proposed a routing metric called ETX [12], which quantifies the average transmission count of an 802.11 link required to have packet received by the other node. Because 802.11 link requires an acknowledgement to complete a transmission, ETX of a link is computed as $ETX = \frac{1}{d_f \times d_r}$, where d_f is the success transmission ratio to the neighbor via the link and d_r is that of the reverse direction. Value d_f are computed via periodically transmitted probe packets with the formula $d_f = \frac{c\tau}{w}$, where τ is the time interval of probe packets, w is the measuring time range, and c is the number of received probe packets within time w. Value d_r is computed in the same way for the reverse direction at the neighbor node, and is carried by probe packets to the neighbor to compute the ETX value of the link.

ETX is extended by introducing communication speed of links as ETT (Expected Transmission Time), which quantifies the average transmission time to have an 802.11 frame received via a link [13]. WCETT (Weighted Cumulative ETT) is also proposed in the same paper [13], which takes bottleneck channel affection into account to compute path metrics under multi-channel environments. Note that WCETT is not link metrics but path metrics ("link metric" here means the additive metric where a path metric is the sum of link metrics included in the path), which is expressed as $WCETT = (1 - \beta)\sum_{i=1}^{n} ETT_i + \beta Max_{1 \le j \le k} X_j$, where n is the number of hops of the forwarding path, k is the number of available channels, and $X_i = \sum_{Hop\ i\ on\ channel\ j} ETT_j$. By including the level of bottleneck channel affection$\beta Max_{1 \le j \le k} X_j$, WCETT quantifies the quality of a whole path.

Unfortunately, this path metric approach may include routing loops even in a static metric situation. For this problem, Sobrinho introduced a necessary and sufficient condition for path metrics to be loop-free in static metric situation, which is called isotonicity [14]. As a path metric that hold isotonicity, Yang et al. proposed a path metric called MIC (Metric of Interference and Channel-switching) [15], which metric values can be decomposed to the isotonic metrics in a virtual network. This characteristic enables MIC to be computed efficiently using the general shortest-path computation algorithms such as Dijkstra's algorithm.

4.3. Loop-prevention for proactive networks

It is now well understood that dynamic metrics significantly improves the network performance and the robustness against instability of wireless links. However, dynamic metrics by nature increase the frequency of changing forwarding paths, which in contrast causes instability of communications in networks. One of the significant effects introduced by dynamic metrics is the routing loop problem described in Section 2. As described there, the routing loop problem is one of the major causes of instability in wireless multi-hop networks.

One naive idea to eliminate routing loops in wireless networks is to apply the loop-free techniques proposed for wired networks. However, with dynamic metrics, it is hard to apply these loop-free techniques such as DUAL because metric change is too frequent. In such cases, diffusing computations such as DUAL requires too many requests for changing successors. Consequently, for wireless multi-hop networks, we need another lower-cost approach against routing loop problems.

As a loop aware routing scheme for MANETs, LLD (Loop-free Link Duration) is proposed [35]. LLD extends proactive routing schemes such as OLSR [3] by introducing dynamic metrics. In LLD, based on the assumption that longer lasting links are more stable in probability, each link metric is decreased constantly as time passes from an initial value. Namely, the longer a link stays stable, the smaller its metric becomes. Every link metric $\delta^t(l)$ of link l at time t is managed by one of the end node of l with the formula $\delta^t(l) = ab^t + c$ as long as the link is judged as "stable, " where t is the time passed by since the link was born, and

b $(0<b<1)$ is a ratio of metric decreased per unit time. When a link is judged as "unstable," the metric is reset with the initial value $\delta^0(l)=a+c$. This judgment should be done carefully because frequent reset of metrics leads instability of forwarding paths.

In LLD routing protocol, link metrics are updated periodically and the timing of them is roughly synchronized in the network. With the synchronization, we can decrease all link metric values in the same ratio, which guarantees loop-freedom as long as no link becomes "unstable." However, not only it cannot ensure loop-freedom when unstable link appears, but also it requires additional messages for synchronization.

As a method to reduce routing loops in wireless mesh networks, LMR (Loop-free Metric Range) is presented in [36]. The idea of LMR is to prevent rapid changes of metrics by applying the changeable range of metrics per unit time. So, LMR can be applied to any dynamic link metric proposed ever. The changeable range in LMR is expressed as the formula $m_{l,t-t'} r^{-t'} \leq m_{l,t} \leq m_{l,t-t'} r^{t'}$, where $m_{l,t}$ is the metric of link l at time t, and r is the coefficient that we call *metric stretch*. Namely, a metric can be changed in a unit time by the value multiplied by metric stretch r.

Note that, reference [36] theoretically proved that we could achieve loop-freedom at any instant with sufficiently small value of r as long as no link fails. Unfortunately, such loop-free value of r is too small to use in practice. Although the loop-free value of metric stretch r depends on several parameters such as network diameter, the value of r to be loop-free is less than 0.5% per unit time for a practical case, where we usually suppose the time interval of control messages in the deployed routing protocol as the unit time. However, reference [36] showed that, even with values r larger than the loop-free threshold, we could reduce routing loops in wireless mesh networks. Here, note that a trade-off is observed that loop packets are reduced for smaller r, but simultaneously the flexibility of paths selection is also reduced so that more congestions and further link failure occur.

5. A new loop-reduction technique for wireless mesh networks

5.1. Idea for loop reduction

In this section, we propose another method to reduce looping in Wireless Mesh Networks that can work in combination with LMR. LMR reduces loop packets by limiting the range of metric changes to prevent rapid transition of metrics. Namely, packet loops are reduced by means of suppressing paths flapping. Here, note that the metric stretch r of LMR that guarantees loop-freedom depends on the network diameter. Specifically, the larger the network diameter in hop count is, the smaller the metric stretch to guarantee loop-freedom is. This means that it is difficult to guarantee loop-freedom for farther destination because larger number of links is included in the path. In other words, packets destined to nearer nodes would less likely to create a loop.

To lower the metric stretch of LMR, it would be reasonable to limit the length in hop count of each single forwarding path. In our method, this is done by splitting a forwarding path between two nodes into the sequence of partial forwarding paths, such that the length of each partial path in hop count is smaller than a certain value k. This is performed using loose source routing, in which a packet should visit several intermediate nodes before reaching its final destination.

By limiting the path length within k, the threshold of the metric stretch of LMR to be loop-free becomes smaller, because we can use the threshold value in which network diameter is assumed to be k. For example, if $k=3$, and the maximum and the minimum link metric are $M_{max}=5$ and $M_{min}=1$, respectively, the metric stretch should hold $r \leq 1.0667$ to guarantee loop-freedom according to [36]. For $k=4$ and $k=5$, the conditions are $r \leq 1.05$ and $r \leq 1.04$, respectively.

5.2. A new loop-reduction method

We propose a new method to reduce routing loops for wireless mesh networks. As described in the previous section, we split a forwarding path between two nodes into a sequence of partial paths to limit the length of each partial path. To achieve this, we introduce a technique of loose source routing, where the list of intermediate nodes to visit is held in each packet header, and the packet is repeatedly forwarded to the next intermediate node to finally reach its destination.

In our method, these intermediate nodes are set along the shortest paths to the destination. Specifically, every time a packet reaches its intermediate node, the node set the next intermediate node with the node *k*-hop ahead in the shortest path to the destination. Note that, with this method, we require only one additional field in the packet header for an intermediate node.

We explain our method with an example shown in Figure 3. There is a network with 10 nodes and assume that the proposed method is working with *k*=2. Suppose that a packet is sent from *A* destined to *F*, and the computed shortest path from *A* to *F* is expressed by the sequence of nodes *A*, *B*, *C*, *D*, *E* and *F*. In *A*, because the intermediate-node field is empty, *A* sets the field with C, which is the 2-hop ahead in the forwarding path to *F*, and forwards the packet to the next-hop node to reach the intermediate node C. Node *B* forwards a packet to C, which is the next-hop node for the intermediate node C. At node C, because C is the intermediate node of the packet, node C update the intermediate-node field with *E*, which is the node 2-hop ahead on the path to the destination *F*, and C forwards packets to *D*. Node *D* simply forwards packet to *E*. At node *E*, because the destination *F* lies within *k* hop distance, it forwards the packet to *F* with its intermediate-node field left empty. The proposed method forwards packets to their destination in this way.

To implement this scheme in practice, nodes are required not only to prepare extra field for an intermediate node in the packet header, but also to maintain an extra table that manages an intermediate node for each destination. Because this scheme extends link-state routing schemes, computing intermediate nodes can be easily done in the process of shortest-paths computation. Consequently, the intermediate-node table, in which the node k-hop ahead on the shortest path for each destination is held, is computed every time a node computes the shortest paths to create its routing table.

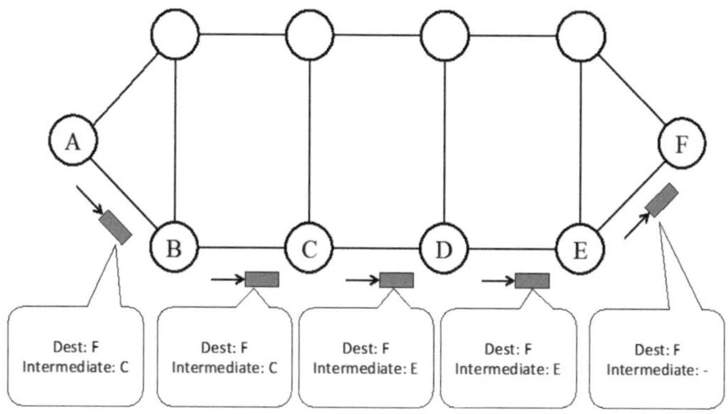

Figure 3. Proposed Method, Case of k=2

5.3. Simulation setup

We evaluated the proposed method through computer simulation with network simulator Qualnet [38]. We implemented ETX link metric and the proposed method by modifying the OLSRv2-Niigata module [39].

As a network topology, we used a 7 x 7 grid as shown in Figure 4, where nodes are placed with interval of 300 meters. Because we set the transmit power with 85dB, only the neighbors in vertical and horizontal directions are connected by wireless links. Each node has a single 802.11 interface with an omni-antenna, and the communication speed is fixed to 2Mbps.

We generated four CBR (Constant Bit Rate) flows in diagonal directions, i.e., from *A* to *D*, *D* to *A*, *B* to *C*, and *C* to *B*, where the packet size is 512 bytes and the communication speeds tried are 40, 60, and 80kbps. We ran the simulation for 15 minutes: during 2 minutes from the beginning, we just wake up all routers and wait for the convergence of routes, and after that we generate the four CBR flows for 10 minutes.

As for OLSR parameters, we use the default values for HELLO_INTERVAL and TC_INTERVAL, i.e., 2 and 5 seconds, respectively. Because we deployed ETX routing metric, we set the parameter NEIGHBOR_HOLD_TIME with 20 seconds, i.e., 10 sequential loss of HELLO messages cause links to fail. We set TC_REDUNDANCY=2 to disable the mechanism of MPR (Multi-Point Relay), which reduce the load of control messages by limiting relay nodes and advertised links, to exclude the affect of advertised link selection. With this parameter settings, all nodes relay control messages and all links are advertised over the network.

We compared the cases of (i) the conventional method only with ETX metric, (ii) the proposed method with ETX metric, and the proposed method with ETX and LMR. As for LMR, we tried several parameters of metric stretch r. We performed 30 trials for each parameter, and compared the performance using the average of them.

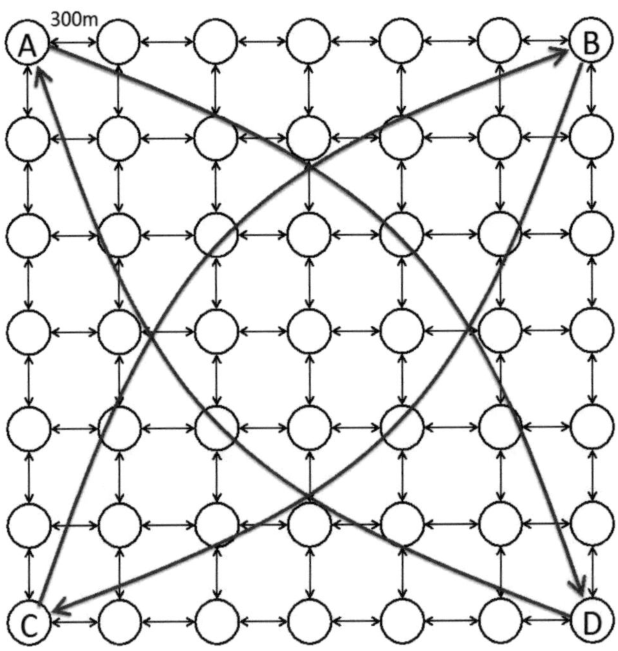

Figure 4. Simulation Settings

5.4. Simulation results

Figures 5-10 show the results of the case where the transmission rates of the four flows are all 40, 60, and 80kbps. In these figure, the results of every combinations of the values k=2, 3, ..., 6 of the proposed method and the metric stretch values r=0.01, 0.03, 0.05, 0.1, 0.2 of LMR.

In Figure 5, the number of loop packets in the 40kbps case is shown. The number of loop packets is relatively low in total, but we see that the ETX cases take especially high value, which surely indicates the effects of the proposed method on reducing loop packets. It is wondering that loop packets increase when the metric stretch takes smaller values. The reason is that; when the metric stretch is small, forwarding paths tend to be persistent and paths selection does not react against rapid transition of link quality, resulting in link failure. Link failure causes paths re-computation and consequently leads packet loops.

On the other hand, in this result, packet loops do not occur frequently when the metric stretch is high or LMR is not applied. This is why the network is not so congested that cause rapid metric changes and then routing loops. For the evidence of this point, see the result of packet delivery ratio shown in Figure 6, where packet delivery ratio is very high as much as 90%, indicating that the congestion level is low.

Figure 6 also shows the interesting tendency that the packet delivery ratio is the best when the metric stretch r takes 1.03. We point out that the main reason of this phenomenon is interference among nodes; when r takes large values, forwarding paths change frequently and it becomes the situation where many nodes have packets to forward. The throughput degrades in CSMA/CA when the number of nodes in contention increases. Namely, when r takes small values, loop packets due to link failure degrades the performance, whereas when r takes large values, heavy interference degrades the performance. The result of Figure 6 shows that the balance point of the metric stretch in this network is around r=1.03.

In Figures 7-10, we show the results in the cases of more loaded scenarios where we generate 60kbps and 80kbps flows. Although the whole tendency is the same as the 40kbps scenario, the number of loop packets increases and the packet delivery ratio decreases in total. Note that in these cases packet loss with the reason "queue over" significantly increases, which indicates the heavy load of the network. Note that, in Figure 9, the cases of k=3, 4, 5 especially take better performance of loop packets reduction. Note that in the cases where r is less than 1.03, routing loops due to link failure dominates, and the cases where both LMR and the proposed method are applied takes better performance.

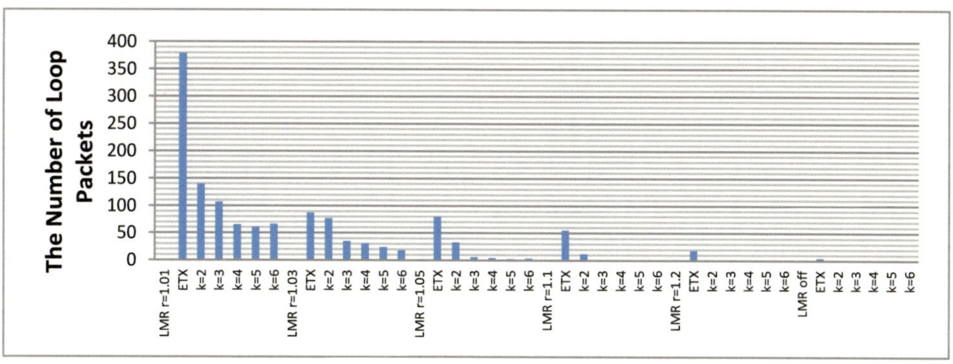

Figure 5. Number of Loop Packets (Case of 40kbps flows)

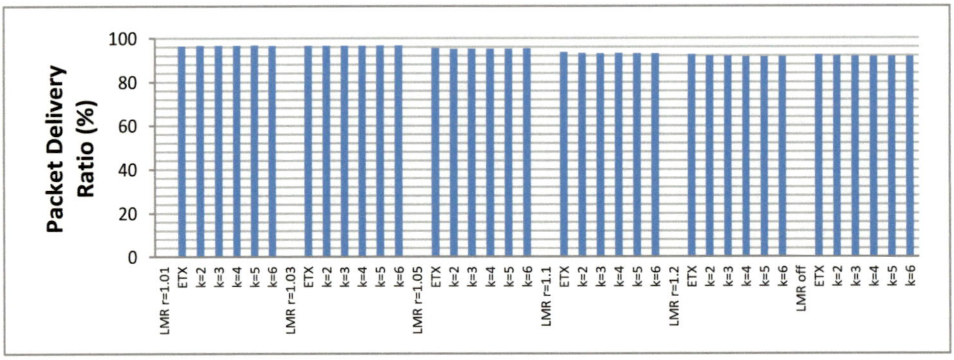

Figure 6. Packet Delivery Ratio (Case of 40kbps flows)

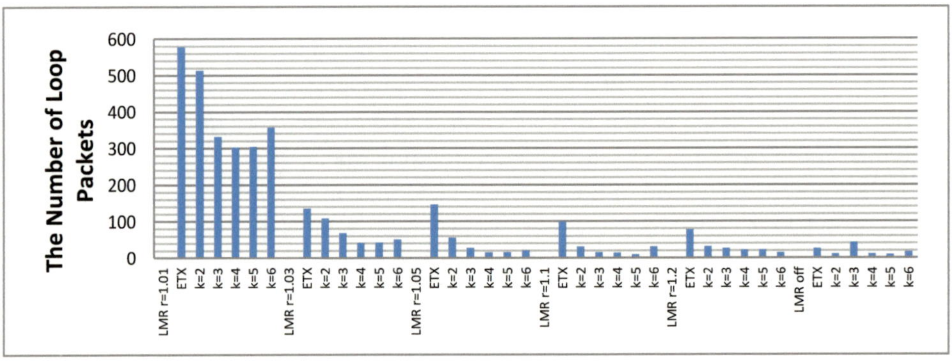

Figure 7. Number of Loop Packets (Case of 60kbps flows)

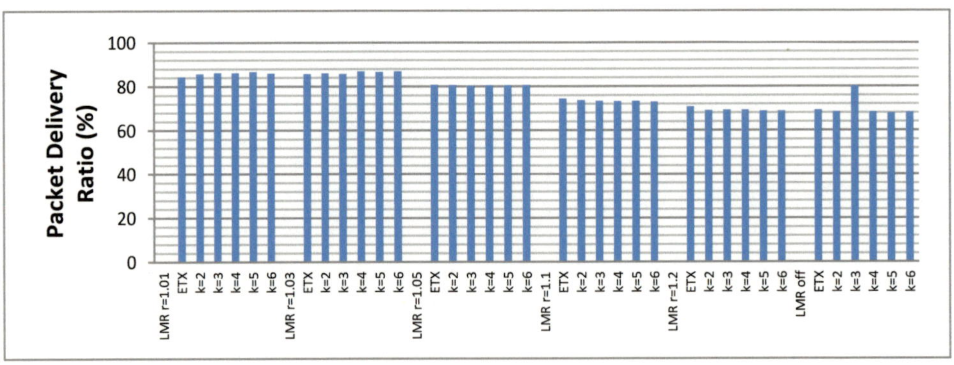

Figure 8. Packet Delivery Ratio (Case of 60kbps flows)

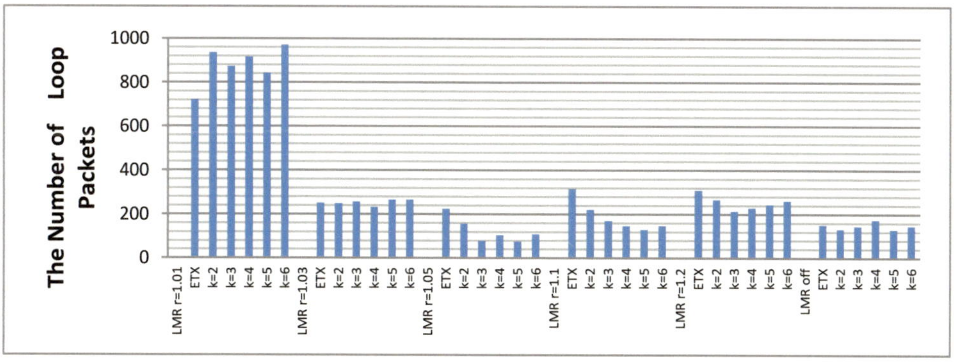

Figure 9. Number of Loop Packets (Case of 80kbps)

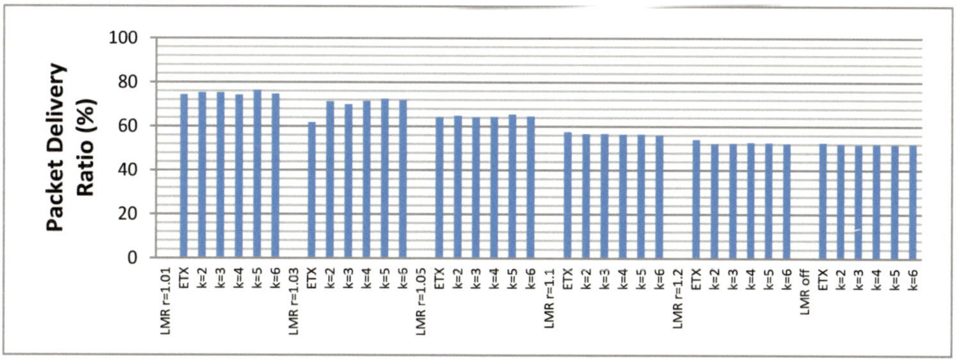

Figure 10. Packet Delivery Ratio (Case of 80kbps)

5.5. Discussion

We evaluated the performance of the proposed method in combination with the other loop reduction method LMR. The simulation results showed that the proposed method solely did not work well and it effectively worked when LMR is applied together. In other words, the combination of the proposed method and LMR works well to reduce routing loops in wireless mesh networks.

Routing loop problem in wired networks has been traditionally regarded as a harmful problem to be avoided, and it is the same in wireless mesh networks. The harmful influence of the looping packets has not been frequently focused on in wireless mesh networks, only because the problem of interference has currently far larger impact on the performance. Although the harmful affection is not always reflected on throughput or packet delivery ratio, looping packets causes large variation of jitter and throughput on time series, which surely degrades the quality and the stability of the communications.

To improve the quality and stability of communications, we have proposed two methods, both of which attain loop-reduction against metric changes. One of the important problems here is that the two methods do not assume link failure, whereas practically congestions easily cause link failure and the consequent packet loops in wireless mesh networks. To make the most of the applied methods and achieve stable communications in wireless mesh networks, it is desirable to develop a congestion control method that prevents link failure even in case of congestion. In combination with such congestion control methods, the harmful influence of looping packets in wireless networks would be significantly improved.

6. Conclusion

In this article, we reviewed the literature of loop-free routing in wired and wireless networks, and further we proposed a new loop-reduction technique for wireless mesh networks. We see that, in proactive routing schemes, it is promising to apply dynamic metrics to afford flexibility against wireless instability, and for this reason it is difficult to apply a family of loop-free techniques developed for wired networks.

We have several loop reduction techniques for wireless multi-hop networks, but it is not still sufficient in performance to provide stable communications in wireless multi-hop networks because loops are not still eliminated. We see that the main cause of looping under the two loop reduction methods, i.e., LMR and the proposed method in this paper, is link failure due to congestion. To provide stable communications over wireless mesh networks, a method is required to prevent link cuts even in case of congestion. To develop such congestion control methods, which works in combination with the two loop-reduction methods, is one of the important tasks to realize wireless mesh networks that can provide stable and reliable communications without routing loops.

Author details

Takuya Yoshihiro and Masanori Kobayashi

Wakayama University, Japan

References

[1] Perkins, C., Belding-Royer, E. & Das, S. (2003), Ad Hoc On-demand Distance Vector (AODV) Routing, IETF Request For Comments (RFC) 3561, IETF.

[2] Johnson, D., Hu, Y. & Malts, D. (2007). The Dynamic Source Routing Protocol (DSR) for Mobile Ad Hoc Networks for IPv4, IETF Request For Comments (RFC) 4728, IETF.

[3] Clausen, T. & Jacquet, P. (2003). Optimized Link State Routing Protocol (OLSR), IETF Request For Comments (RFC 3626), IETF.

[4] Ogier, R., Templin, F. & Lewis, M., (2004). Topology Dissemination Based on Reverse-Path Forwarding (TBRPF), IETF Request For Comments (RFC 3684), IETF.

[5] Akyildiz, I.F. & Wang, X. (2009), Wireless Mesh Networks, John Wiley & Sons Ltd Publication.

[6] Cheng, Z. & Heinzelman, W.B. (2004). Exploring Long Lifetime Routing (LLR) in ad hoc Networks, Proceedings of 7th ACM International Symposium on Modeling, Analysis and imulation of Wireless and Mobile Systems, pp.203–210.

[7] Zhao, M. & Wang, W. (2007). The impacts of radio channels and node mobility on link statistics in mobile ad hoc networks, Proceedings of IEEE Global Telecommunications Conference (Globecom2007), No.1, pp.1205-1209.

[8] Tickoo, O., Raghunath, S. & Kalyanaraman, S. (2003). Route Fragility: A Novel Metric for Route Selection in Mobile Ad Hoc Networks, Proceedings of IEEE ICON03, pp. 537–542.

[9] Trivi ˜no-Cabrera, A., Nieves-P′erez, I., Casilari, E. & Gonz′alez-Ca˜nete, F.J. (2006). Ad Hoc Routing Based on the Stability of Routes, Proceedings of the 4th ACMInternationalWorkshop on MobilityManagement and Wireless Access, pp.100–103.

[10] Yawut, C., Paillassa, B. & Dhaou, R. (2007). On Metrics for Mobility Oriented Self Adaptive Protocols, Proceedings of in Wireless and Mobile Communications 2007 (ICWMC2007).

[11] Qin, L. & Kunz, T. (2006). Mobility Metrics to Enable Adaptive Routing in MANET, Proceedings of IEEE Wireless and Mobile Computing, Networking and Communications 2006 (WiMobapos2006), pp.1–8.

[12] De Couto, D., Aguayo, D., Bicket, J. & Morris, R. (2003). A High-Throughput Path Metric for Multi-Hop Wireless Routing, Proceeding of ACM Annual International Conference on Mobile Computing and Networks (MOBICOM2003), pp.134–146.

[13] Draves, R., Padhye, J. & Zill, B. (2004). Routing in Multi-Radio, Multi-HopWireless-Mesh Networks, Proceedings of ACM Annual International Conference on Mobile Computing and Networks (MOBICOM2004), pp.114–128.

[14] Sobrinho, J. L. (2003). Network Routing with Path Vector Protocols: Theory and Applications, Proceedings of the ACM 2003 conference on Applications, technologies, architectures, and protocols for computer communications (SIGCOMM2003), pp.49–60.

[15] Yang, Y., Wang. J. & Kravets, R. (2005). Interference-aware Load Balancing for Multihop Wireless Networks, In Technical Report UIUCDCS-R-2005-2526, Department of Computer Science, University of Illinois.

[16] Jin, S., & Mase K., A Hidden-Exposed Terminal Interference Aware Routing Metric for Multi-Radio and Multi-Rate Wireless Mesh Networks, IEICE Transactions on Communications, Vol.92-B(No.4), pp.709–716.

[17] Speakman L., Owada Y., Mase K., Looping in OLSRv2 in Mobile Ad-Hoc Networks, Loop Suppression and Loop Correction, IEICE Transactions on Communications, Vol.E92-B(No.4), pp.1210–1221.

[18] A.S. Tanenbaum, Computer Networks Fourth Edition, Publisher: Prentice Hall PTR, ISBN:9780130661029, 2003.

[19] G. Malkin, "RIP Version 2," IETF RFC2453, November 1998.

[20] J.M. Jaffe and F.M. Moss, "A Responsive Routing Algorithm for Computer Networks," in IEEE Transactions on Communications, Vol. COM-30, No. 7, July 1982, pp. 1758-1762.

[21] Y. Rekhter, T. Li, and S. Hares, "A Border Gateway Protocol 4 (BGP-4)," IETF RFC 4271, January 2006.

[22] C. Cheng, R. Riley, S. P. R. Kumar, J. J. Garcia-Luna-Aceves, A loop-free extended Bellman-Ford routing protocol without bouncing effect, SIGCOMM Comput. Commun. Rev., Vol. 19, No. 4. (September 1989), pp. 224-236.

[23] Garcia-Luna-Aceves, J. J. (1993), Loop-free Routing using Diffusing Computations, IEEE/ACM Transactions on Networking, Vol.1 (No.1), 130–141.

[24] R. Albrightson, J.J. Garcia-Luna-Aceves, and J. Boyle, "EIGRP – A Fast Routing Protocol Based on Distance Vectors," in Proceedings of Network/Interop, Las Vegas, NV, May 1994.

[25] J.J. Garcia-Luna-Aceves and Shree Murthy, A Path-Finding Algorithm for Loop-Free Routing, IEEE/ACM Trans. Networking, Vol.5, pp.148—160, 1997.

[26] Andreas Schmid and Christoph Steigner, Avoiding Counting to Infinity in Distance Vector Routing, TELECOMMUNICATION SYSTEMS, Volume 19, Numbers 3-4 (2002), 497-514, DOI: 10.1023/A:1013858909535

[27] IS-IS (Intermediate Systems – Intermediate Systems), ISO/IEC 10589, 2002.

[28] Moy, "OSPF version 2," IETF RFC2328, April 1998.

[29] Francois, P. & Bonaventure, O. (2007), Avoiding Transient Loops During the Convergence of Link-state Routing Protocols, IEEE/ACM Transactions on Networking, Vol. 15 (No. 6), 1280–1932.

[30] Francois, P., Shand, M. & Bonaventure, O. (2007). Disruption-free Topology Reconfiguration in OSPF Networks, Proceedings of IEEE INFOCOM2007.

[31] Ito, H., Iwama, K., Okabe, Y. & Yoshihiro, T. (2003). Avoiding Routing Loops on the Internet, Theory of Computing Systems, Vol.36, 597–609.

[32] J. Raju and J. J. Garcia-Luna-Aceves, "A New Approach to On-demand Loop-free Multipath Routing," In IEEE IC3N'99, pp.522-7, 1999.

[33] J. J. Garcia-Luna-Aceves, "A New Approach to On-demand Loop-free Routing in Ad Hoc Networks," In Proceedings of the twenty-second annual symposium on Principles of distributed computing (PODC'03), pp. 53 – 62, 2003.

[34] Yoshihiro, T. (2009). Loop-free Link Stability Metrics for Proactive Routing in Wireless Ad Hoc Networks, Proceedings of IEEE ICC2009, pp.1–5.

[35] Takuya Yoshihiro (2011). LLD: Loop-free Link Stability Metrics for Proactive Link-State Routing in Wireless Ad Hoc Networks, Mobile Ad-Hoc Networks: Protocol Design, Xin Wang (Ed.), ISBN: 978-953-307-402-3, InTech, Available from: http://www.intechopen.com/books/mobile-ad-hoc-networks-protocol-design/lld-loop-free-link-stability-metrics-for-proactive-link-state-routing-in-wireless-ad-hoc-networks

[36] T. Yoshihiro, "Reducing Routing Loops under Dynamic Metrics in Wireless Mesh Networks," IEEE Global Communication Conference, Exhibition, and Industory Forum (Globecom2010), pp.1-6, 2010.

[37] Qualnet, http://www.scalable-networks.com/

[38] OLSRv2-Niigata, http://www2.net.ie.niigata-u.ac.jp/nOLSRv2/olsrv2/Welcome.html

Section 3

Applications of Wireless Ad-Hoc Networks

Chapter 6

Review of Autoconfiguration for MANETs

Hongbo Zhou and Matt W. Mutka

Additional information is available at the end of the chapter

http://dx.doi.org/10.5772/52687

1. Introduction

A MANET is a temporary multi-hop wireless network composed of mobile nodes without an underlying infrastructure. The mobile nodes in a MANET are not connected to an access point to access the Internet (although in some special cases a MANET and the Internet may co-exist). Instead, their wireless network interface cards operate in ad-hoc mode. The nodes that are within the transmission range of each other can communicate directly. For the nodes that are out of the range, they have to resort to the nodes in between to relay the messages.

Due to the popularity of mobile devices and independence from the infrastructure, a MANET can find wide applications in temporary wireless networks in meeting rooms, airports, and stadiums. It is fast, convenient, and economical to set up a MANET in a battlefield and for search and rescue. A Vehicular Ad-hoc Network, an variation of MANET, connects the running cars and fixed traffic lights and other sensors, is vital to implementation of smart transportation.

Before the application of such an IP-based network, IP address assignment is one of the most important network configuration parameters for the mobile nodes. Without a valid unique IP address, a mobile node cannot participate in unicast communications. It can only receive and send broadcast messages, which consumes valuable bandwidth and power, and thus it is desirable to limit the duration and scope of broadcast communications in a MANET.

For a small-scale MANET or a closed MANET, it may be simple to assign IP addresses to mobiles node by hand. It is also possible to burn an IP address in the ROM of a mobile node to re-use it repeatedly. However, the procedure will become inefficient and even impractical for a large-scale system or an open system where different kinds of nodes (such as laptops, smartphones, tablets, PDAs, and specialized computers) are free to join and leave.

Automatic IP address allocation is far more difficult to implement in a MANET than a hard-wired network such as a local area network, due to instability of mobile nodes, multi-hop transmission of messages, openness of the system, and lack of infrastructure. Therefore, although DHCP [1] or SAA [2] is popular for hardwired networks, they cannot be directly ported to a MANET. A distributed algorithm that adapts to node mobility and topology change is more desirable.

Autoconfiguration for MANETs brings other issues that need to addressed as well.

The first issue is the change of IP address during communication. This issue is rare for a fixed node in a hardwired network running DHCP service. Even after the lease expires, the client tends to receive the same IP address from the server. In a MANET with autoconfiguration, there may be address conflicts when two separate MANETs join together. As a result, some of the nodes need to relinquish the duplicate addresses, which will interrupt on-going communications. An IP address handoff mechanism is necessary to make the transition smooth.

Security of the autoconfiguration is another issue that needs to be considered. If malicious nodes may be present, they may launch attacks on the autoconfiguration scheme to make it fail. Thus, a secure autoconfiguration is necessary for applications that have strict requirement for security level.

Autoconfiguration impacts the design of the security framework in a MANET. Most security frameworks assume that the IP address is the ID of the node, which is associated with the node's security parameters such as its public key. In an autoconfiguration scheme, the IP address is generated dynamically, and it may need to change during the session, which exposes the vulnerabilities in some pre-existing security frameworks.

This chapter gives a comprehensive review of autoconfiguration and related issues. It is structured as follows. Section 2 gives a review of autoconfiguration schemes, which can be divided into three groups. Among others, our Prophet Address Allocation outperforms in terms of bandwidth and latency. Section 3 introduces the IP address handoff scheme, which maintains the routing fabrics and on-going communications if there is an address change. In Section 4, we present different attack patterns and improve Prophet Address Allocation to survive these attacks. Thus, unique IP address allocation can still be achieved in the presence of malicious nodes. In Section 5, the Sybil attack is demonstrated to defeat some security frameworks within the scenario of autoconfiguration, and thus a different security framework is desirable. To further solve this problem, a combination of secure autoconfiguration and security framework is described in Section 6. Section 7 concludes the chapter.

2. Autoconfiguration for MANETs

Because unique IP address allocation is the first step towards a functioning MANET, this section studies the issue of autoconfiguration itself.

2.1. Introduction

In an open MANET where mobile nodes are free to join and leave, there are three scenarios that are of concern:

a. In the simplest scenario in Fig. 1, the node joins the network and then participates in the communication with other nodes. Once it is done, it leaves the network forever;

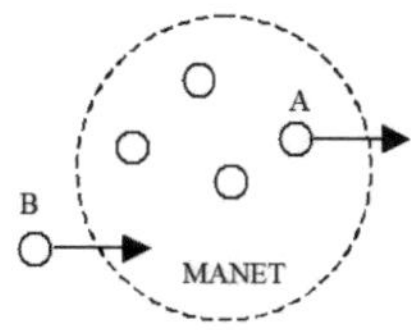

Figure 1. A node joins and leaves the MANET once

b. It is possible that some nodes move out of transmission range of other nodes. Thus, the network is divided into two or more partitions, as in Fig. 2 (a). If more nodes join different partitions, they may get duplicate addresses. When partitions merge later, address conflicts need to be resolved, as in Fig. 2 (b);

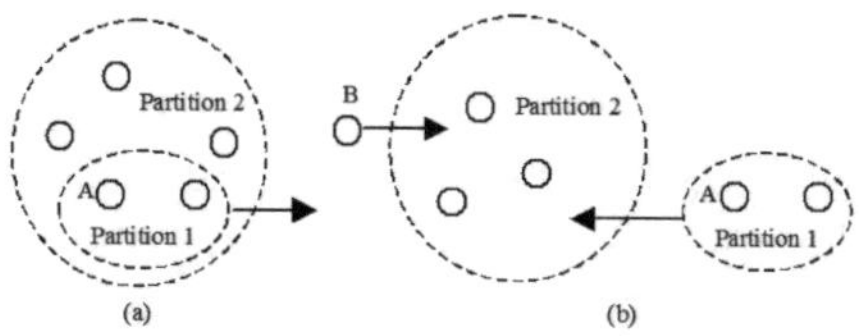

Figure 2. Network partitions and merges

c. In the third case, two or more independent MANETs merge. Because IP addresses are allocated independently, there may be duplicate addresses, which is similar to the second case.

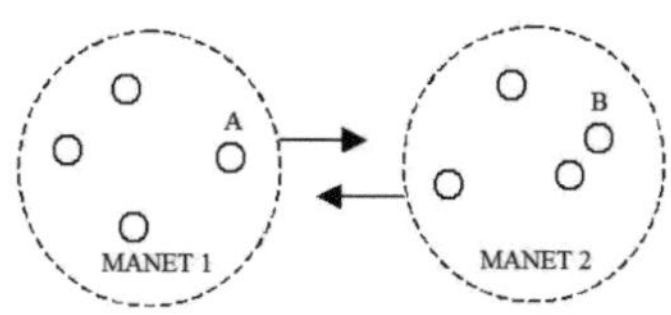

Figure 3. Merger of two independent MANETs

The autoconfiguration scheme needs to address all the three scenarios.

2.2. Related work

Several autoconfiguration schemes have been proposed. Depending on the IP address allocation state they have, they can be divided into the following three groups:

a. Stateless allocation

In the stateless allocation algorithm, no IP address allocation state is recorded. Instead, it utilizes Duplicate Address Detection (DAD) to determine a free IP address for a newcomer, as in [3]. The scheme reserved a pool of IP addresses for DAD communication only. The new node chooses one IP address from the pool to broadcast DAD messages, which contains another random IP address for actual data communications. The broadcast is utilized to see if the address is still available. If not, the new node will receive a veto message from some member and then choose another random address again. Otherwise, it repeats the DAD procedure for a few more times. Once it is certain that the address is free, it can proceed with the address for subsequent data communications.

The scheme in [3] did not address scenarios 2 and 3. A modified version, which is called weak DAD, was proposed in [4] to detect duplicate addresses on network merger. It favors proactive routing protocols and requires some changes to routing protocols.

b. Locally stateful allocation

The locally stateful allocation, such as one in [5] is based on a buddy system. Each allocator maintains a disjoint address pool. When a new node joins the network, the allocator divides the address pool into halves between itself and the new node. Thus, the address received from the allocator is guaranteed to be unique.

Although this algorithm can solve the scenarios 1 and 2 easily, it lacks the mechanism for scenario 3. Another problem is the reclamation of lost address pools. To solve this problem, allocators may need to exchange beacon messages frequently to track all the address pools and thus result in high communication overhead.

A similar idea was proposed in [6] that tries to deal with the network's partition and merger.

c. Globally stateful allocation

Distributed Dynamic Host Configuration Protocol (DDHCP) proposed in [7] maintains the global allocation state of IP addresses in all the members. When a new node joins, the allocator chooses a free IP address according to the allocation state. But the new node still needs to perform DAD to avoid the allocation of the same free IP address to two new nodes arriving simultaneously. It is also used to update the global allocation state in all the members.

DDHCP works well with proactive routing protocols. It also introduced network ID to detect merger of partitions or independent networks. It is generated from the node with the lowest IP address and is piggybacked in the periodic HELLO messages. Once network merger is detected, conflict detection and resolution will be initiated.

2.3. Prophet Address Allocation

We proposed Prophet Address Allocation (PAA) in [8] and [9], in which each node maintains a local allocation state. Unlike the disjoint address pools used in [5], the allocation state in PAA is an integer sequence. The integer sequence is generated by a stateful function with a seed. We deliberately designed the stateful function and update of seeds to satisfy the following two properties:

i. The interval between two occurrences of the same integer in a given sequence is extremely long;

ii. The probability of the same number in different sequences with different seeds in a given allocation session is extremely low.

The general procedures work as below:

iii. The first node chooses an initial state. Based on the initial state, it generates an integer as its own IP address. It updates the state;

iv. When another node joins the MANET, it receives the new state from the allocator, which is used to calculate its own IP address. The states in the allocator and the new node are updated simultaneously.

The design of the stateful function is based on the fundamental theory in arithmetic that every positive integer may be expressed uniquely as a product of prime numbers, apart from the arrangement of terms, as in the following formula:

$n=\prod_{i=1}^{k} pi^{ei}$, where the primes pi satisfy p1<p2< ... <pk.

If each node has different tuples, they will have unique integers for their IP addresses. Thus, the most important issue is to generate different tuples during allocation. In our design, we chose the state to be (a, (p1, p2, ...,pk), index), and the stateful function is:

$$IPAddr = (a+\prod_{i=1}^{k} pi^{ei}) \ \% \ range + 1$$

Parameter a in the state is used by the first node in each MANET to generate unique seed. The tuple is used to generate the product of prime numbers. Parameter index is the location of exponential value in the tuple that is to be increased by the allocator, while index itself is increased at the new node.

Based on the state and stateful function as above, the procedures of PAA can be illustrated with a 4-tuple:

a. The first node chooses a random number for a, an initial tuple of (0, 0, 0, 0) for p1 to p4, and 0 for index. It generates its own IP address, which is a + 1;

b. When the second node approaches the first node, the allocator updates its state to (a, (1,0, 0, 0), 0), and passes it to the new node. The new node increases index by 1. Thus, the new node's state is (a, (1,0, 0, 0), 1), and its address is a + 3;

c. If the third node approaches the first node, the first node updates its state to (a, (2,0, 0, 0), 0) and passes it to the new node. The third node gets the address of a + 5, and updates its own state to (a, (2,0, 0, 0), 1).

d. Similarly, when the fourth node approaches the second node, the allocator updates its state to (a, (1,1, ..., 0), 1) and passes it to the third node. The new node increases index by 1. Thus, the third node's state is (a, (1,1, 0, 0), 2), and its address is a + 7.

In summary, all the four nodes will have different states and different IP addresses. They will generate different integers in a sequence in the subsequent allocations.

2.4. Protocol

The protocol of the PAA includes the following steps:

i. When a new node switches to ad-hoc mode, it starts to broadcast State Request Message periodically. Because each node in the MANET can act as an allocator, so only one-hop broadcast is necessary;

ii. If the new node does not receive any State Reply Message, it concludes that it is the first node in the MANET. Thus, it generates a random number for a and configures itself;

iii. If it receives a State Reply Message that contains the state, it applies the stateful function to configure itself, and update the state;

iv. During its session, if it receives a State Request Message from some other new node, it updates its state and includes it in the State Reply Message.

The above-mentioned procedures can solve Scenario 2 easily because even the network is partitioned, the addresses allocated in different parts will still be different. To handle Scenario 3, we borrowed the idea of Network ID (NID) from DDHCP. The first node also chooses a random NID and propagate it throughout the network during the allocation. If seed a and NID are also contained in the periodic HELLO messages, the nodes between two separate MANETs will detect the merger of two networks and initiate conflict resolution procedures. With the seeds, the node on the border can calculate two integer sequences and locate potentially duplicate addresses. The duplicate address list is then broadcast throughout the two MANETs. If a node happens to have that duplicate address, it changes its address accordingly.

2.5. Performance evaluation

Because every node can be an allocator in Prophet Address Allocation, only the communications between one-hop neighbors are necessary. Thus, PAA outperforms other schemes in the term of bandwidth and latency and is more suitable for a large-scale MANET.

We also ran simulations to demonstrate its superiority of PAA over stateless address allocation with ns-2. The simulations are run on ns-2.34 [10] with 50 nodes to 250 nodes. The random waypoint mobility model was used in the simulation [10]. After a node pauses for several seconds, a random destination point is chosen. The maximum speed is set to 5 m/s, which is repeated until the end of simulation. The pause time is 10 s for 50 nodes and 100 nodes, 20 s for 150, 200, and 250 nodes. These nodes join the MANET every 30 s (for 50, 100, and 150 nodes), or every 10 s (for 200 and 250 nodes). Different area sizes are also introduced to show the effect of node density on the algorithm. For example, scenario files of 800×800, 1000×1000, and 1200×1200 are utilized for 100 and 150 nodes, while scenario files of

1000×1000, 1200×1200, and 1400×1400 are tested for 200 and 250 nodes. The final results are the average results obtained with all the area sizes.

Although we chose AODV as the ad hoc routing protocol during the simulation, both address allocation schemes use one-hop and multi-hop broadcast, respectively for control message exchanges.

Figure 4 shows the ratio of total messages collected in stateless address allocation and Prophet Address Allocation, in contrast with a linear line. It shows that the ratio increases linearly with the number of nodes. Because in stateful allocation, each node is going to receive one copy of the message from all its neighbors, while in PAA only the neighbors of the new node receive a copy, so the communication overhead is almost constant regardless of the number of nodes in the MANET.

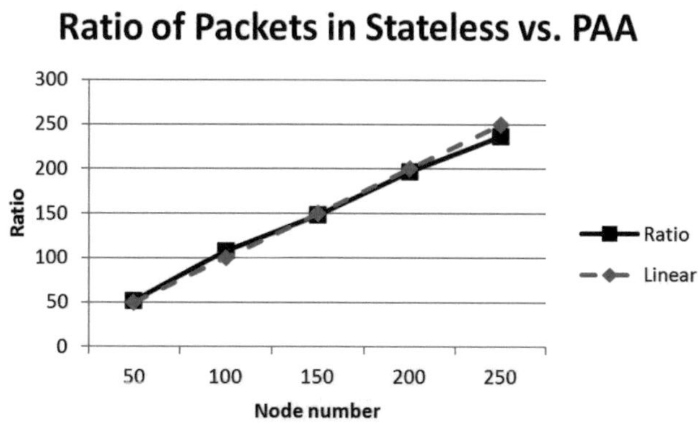

Figure 4. Ratio of communication overhead of Stateless to PA

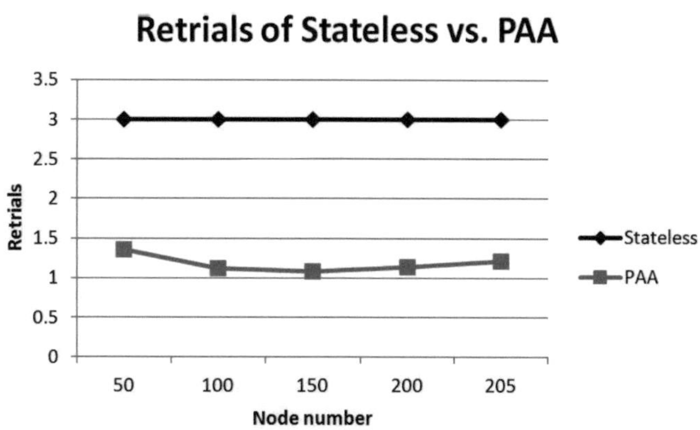

Figure 5. Latency for different node numbers

Figure 5 shows the average number of retrials for all the nodes in the MANET. For stateless allocation, each node tries for a constant three times. For APP, except for the first node that tries for three times, all the nodes try for infinite time. However, the simulation results show that most of them get reply messages within 2 rounds.

3. IP address handoff in MANETs

With autoconfiguration implemented in MANETs, the IP address of a mobile node is not fixed any more. If a node changes its IP address, on-going communications will be interrupted, and routing fabrics will be broken. Thus, we need a handoff scheme to address these issues.

3.1. Introduction

Due to the node mobility and topology change, the following scenarios may lead to the necessity of IP address change in some mobile nodes:

a. A MANET is divided into two or more partitions. Some new members join different partitions and duplicate addresses may be allocated. Once the partitions merge, these nodes need to change their IP addresses;

b. Two separate MANETs merge together. Because address allocation is independent in each network, there could be duplicate addresses;

c. In case where hierarchical addressing scheme is applied [11], a node needs to change its address when roaming from one subnet to another, as in Fig 6;

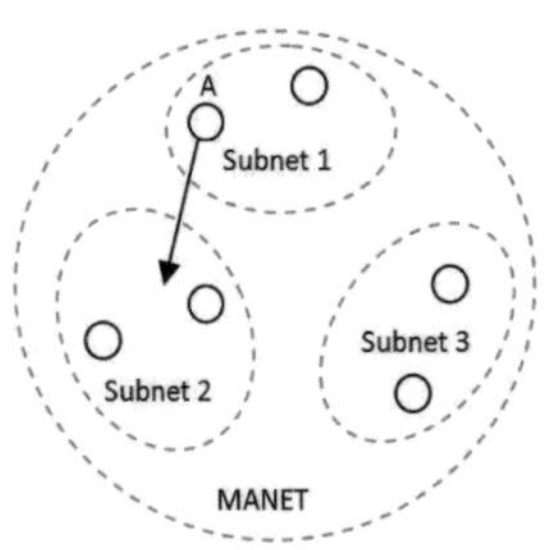

Figure 6. A MANET with hierarchical addressing

d. A MANET can get connected to the Internet if a hardwired node that also has a wireless network interface card working in ad-hoc mode, as in Fig. 7. That node would behave like a bridge between the hardwired network and the MANET. If the MANET and the hardwired network use the same private address range, there could be duplicate addresses.

When a node (say node A) changes its IP address, its current data communication will be interrupted. Even the other party (say node B) may initiate to rebuild the connection, it will not find node A because that IP address does not exist anymore, unless the DNS scheme proposed in [12] is combined with the reactive routing protocol.

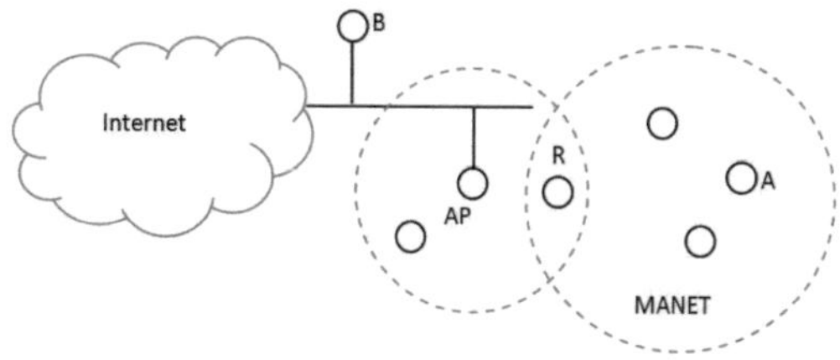

Figure 7. A MANET is connected to a hardwired network

Another issue with IP address change is the routing fabrics will be broken. Although the effect is not immediate since the forwarding is based on next hop's MAC address. The ARP entry will gradually time out and its upstream nodes cannot reach it anymore. Thus, local routing repair mechanism or route rediscovery procedures needs to be initiated eventually.

IP address change will also pose the threat to privacy. Suppose node A is talking with node B. On merger of two MANETs, node A needs to change its address because it has the same address of node C. Without notification to node B, the messages from node B will be routed to node C.

3.2. Related work

Mobile IP supports host mobility among LANs [13]. The host is assigned with a permanent home address. Once it connects a foreign network and gets a temporary care-of address, it registers the care-of address at its home agent. A tunnel is then built to forward messages between its home agent and foreign agent. Thus, any messages that destined to its home address will be forwarded to its current care-of address. Because a MANET is different from a LAN, so it is not working for address handoff.

The tunneling scheme was proposed in [14] that aimed to maintain on-going communications after address change in MANET. Once a node changes its address, it sends an Address Error (AERR) to the other party and an IP-in-IP tunnel is created between them. The outer IP header contains the new address while the inner IP header uses the old address. This solution can preserver communication states. However, it ignores the overhead cause by routing repair or route rediscovery. Besides, it introduced Denial of Service issue, as illustrated in Fig. 8. Suppose node A changes its address from xto y because the another node (node C) has the same address of x. A tunnel is created between nodes A and B that forwards all the packets destined to IP address x to node A. Thus, node B cannot communicate with node C anymore.

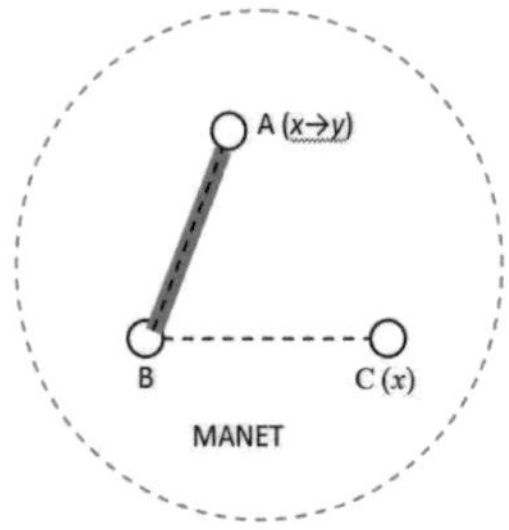

Figure 8. Denial of Service issue in tunneling scheme

3.3. IP address handoff scheme

We proposed IP address handoff scheme in [15] to maintain routing fabrics and keep ongoing communications.

Firstly, because the node is aware of the IP address change, we can require that node broadcast a Route Shift message that contains its old and new IP addresses to its one-hop neighbors. To prevent IP spoofing attacks, we can also require the message to be signed with its private key. A lightweight solution is that the node chooses a random number and attaches the hash value of the random number in Route Request message, Route Reply message, and periodic HELLO messages. In the Route Shift message, the random value is included to verify the identity of the origin.

Secondly, a NAT-based solution is utilized to maintain communication states, in which the old address is changed to new address for outgoing packets while the new address is changed back for incoming packets. NAT needs to be performed at both ends, as below:

a. On address change, node A as in Fig 7 creates a NAT table (such as Table 1) that maps the old address to the new address according to the protocol and source port number of outgoing TCP/UDP packets if the packet still contains the old address.

Old Address	New Address	Protocol	Port Number
x	y	TCP	5472
x	y	...	...

Table 1. NAT table at node A

According to the NAT table, the outgoing packet's old source address x is modified to y, and the checksums need to be re-calculated. For incoming packets, if its protocol and destination port number match an entry, the destination address is changed back to x together with the checksums updated.

b. Node A sends an Address Change message to node B so node B can create its NAT table. The control message contains the old address, new address, node A's port number, node B's port number, and sequence number of message from A to B. Based on these data, node B can insert an entry to its NAT table, such as table 2:

Old Address	New Address	Protocol	Remote Port Number	Local Port Number	Seq Number
x	y	TCP	5472	80	626473
x	y	...	...	...	...

Table 2. NAT table at node B

If an incoming packet matches an entry in the NAT table, its new address y is changed to x. At the same time, the Seq Number field is increased by the payload length. For an outgoing packet with the destination address x, if it matches an entry, its destination address will be set to y. The NAT table at node B includes both port numbers and sequence number from node A, so the packets destined to node A can be distinguished from those to node C. To prevent IP spoofing attacks, the Address Change message is required to include node A's digital signature.

3.4. Prototype implementation

A prototype of IP address handoff scheme was implemented on Fedora Linux 12 with netfilter [16] to test the preservation of communication states in a LAN, as illustrated in Fig. 9. After a TCP connection is created between a laptop client and a server, the client changes its address from 192.168.1.155 to 192.168.1.140.

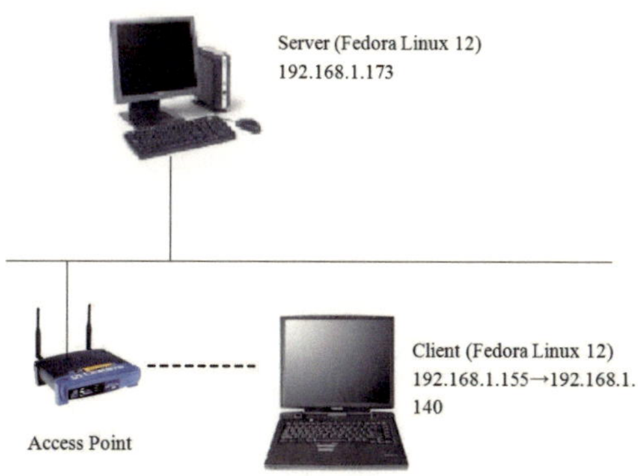

Figure 9. Thetestbed of the prototype

The outgoing packets are handled in NF_IP_LOCAL_OUT hook, while the incoming packets are modified in NF_IP_PRE_ROUTING hook. The code to modify outgoing packets is the client is illustrated in Table 3. The processing of incoming packets is similar. In the prototype, the addresses are hardcoded. But in real application, NAT tables mentioned in the previous section should be utilized.

```
#define OLD_ADDRESS 0xC0A8019B // 192.168.1.155
#define NEW_ADDRESS 0xC0A8018C // 192.168.1.140
static unsigned int handoff_NAT_out(unsigned int hook, struct sk_buff **pskb, const struct net_device *indev,
const struct net_device *outdev, int (*okfn)(struct sk_buff *)
{
struct tcphdr* th;
// Whether we should perform NAT or not
if ((*pskb)-"/>nh.iph-"/>saddr == htonl(OLD_ADDRESS) && (*pskb)-"/>nh.iph-"/>protocol == 6)
{
th = (struct tcphdr*)((char*)(*pskb)-"/>nh.iph + (*pskb)-"/>nh.iph-"/>ihl*4);
if (th-"/>dest == htons(10000))
{
// Change (source) address
(*pskb)-"/>nh.iph-"/>saddr = (DWORD)htonl(NEW_ADDRESS);
// Recompute IP checksum
(*pskb)-"/>nh.iph-"/>check = 0;
(*pskb)-"/>nh.iph-"/>check = in_checksum((WORD*)((*pskb)-"/>nh.iph), (*pskb)-"/>nh.iph-"/>ihl*4);
}
}
return NF_ACCEPT;
}
```

Table 3. NAT processing of outgoing packets at the client

We ran Wireshark at both the server and client to capture all the TCP packets and verified that the session continues without being aware of the address change.

4. Secure address allocation for MANETs

All the autoconfiguration schemes introduced in Section 2 assume that every node in the MANET is trustworthy and have no security mechanism. Thus, if there is a malicious node in the network, the autoconfiguration scheme may fail: either no new node will be allowed to join the network, or there will be duplicate addresses. This section focuses on a secure autoconfiguration scheme.

4.1. Attacks on autoconfiguration

There are several common kinds of attacks that target at autoconfiguration schemes:

a. IP spoofing attack

IP spoofing attack means the malicious node impersonates as another node. In stateless autoconfiguration scheme, the malicious node masquerades as a node with the same IP address as the new node chooses. For each request message from the new node, it replies with a veto message to deter the new node from joining the network. In Prophet Address Allocation, the malicious node uses another member's IP address and state in allocation, and thus duplicate addresses will be assigned.

IP spoofing attack is extremely difficult to detect and prevent, even with the deployment of Certificate Authority, because the IP address of CA itself needs to be obtained with autoconfiguration.

b. State pollution attack

In stateful allocation schemes, the malicious node passes incorrect or forged state in the reply message to the new node, thus duplicate addresses will be allocated.

c. Sybil attack

In Sybil attack [17], a malicious node impersonate several non-existent nodes at the same time. Thus, these nodes will seem like a group and can undermine the network service that requires cooperation among all or most nodes. In stateful allocation scheme, a malicious node can initiate a Sybil attack and convince the new node of incorrect parameters.

4.2. Secure prophet address allocation

In the original Prophet Address Allocation, the new node does not verify the parameters in the reply message, which leaves the exposure to IP spoofing attacks and state pollution attacks. In Secure Prophet Address Allocation, we made the following improvements to guarantee a unique state for each address allocation:

a. More parameters are included in the reply message

Now the reply messages contains the following parameters: (1) The seed value of the MANET (a); (2) The index of increasing exponential (c); (3) The source address of the allocator (x); (4) The initial exponential array (i[1,2, …, n]); (5) Priority (p), which indicates the freshness of the parameters.

b. The new node verifies the parameters and chooses a seemingly correct one randomly

The relationship among these parameters is:

$$x = f\left(a, i\left[1, 2, \ldots, n\right]\right) \tag{1}$$

which means the source address of the allocator should be calculated with the seed value and the initial exponential array.

The new node chooses the reply message randomly that confirms to the relationship and has the highest p value, chooses a random value (r), and generates its own address with the following equation:

$$y = f\left(a, e\left[1,2,\ldots,n\right]\right) \tag{2}$$

where

$$e\left[j\right] = \begin{cases} i[j], & j<c \\ r+p, & j=c \\ i[j]=0, & j>c \end{cases}$$

c. A broadcast is included to force all the members to update its state

An ACK message is broadcast throughout the MANET to include p + r value, which will be the new priority value in each member's state.

Although all the parameters in a reply message could be forged, since the new node always chooses the highest priority value and its own random value, it is guaranteed that the state it receives would be unique at present. The subsequent ACK message forces all the members to update their states. Thus, the state that the new node just used will become obsolete and will not be used in the subsequent allocations.

4.3. Simulation

Simulations of Prophet Address Allocation and Secure Prophet Address Allocation in the are run with ns-2 (version 2.34). Malicious nodes are randomly chosen during the simulation. Statistics data about the number of duplicate addresses are collected to show the invulnerability of the latter. Both allocation schemes are tested with 50 nodes in the area of 800 × 800 with random waypoint mobility model. The size of exponential array is chosen to be 50. However, at most 6 exponentials are used. Table 4 shows that no duplicate addresses are generated in Secure Prophet Address Allocation.

Percentage of malicious nodes	IP spoofing attacks		State pollution attacks and Sybil attacks	
	PA	SPA	PA	SPA
10%	1	0	1	0
20%	2	0	2	0
25%	2	0	2	0
33%	5	0	5	0
50%	3	0	3	0

Table 4. The number of duplicate address in simulations

http://dx.doi.org/10.5772/52687

5. Autoconfiguration's impact on distributed certificate authority

Most prevalent Distributed Certificate Authority (DCA) schemes for MANET are based on threshold cryptography. However, in the context of autoconfiguration, these DCA schemes will fail in the presence of Sybil attacks. This section introduces a different DCA scheme.

5.1. Threshold cryptography-based DCA

Threshold cryptography-based DCA scheme [20] [21] was originally proposed for hard-wired networks, in which the administrators of servers can verify others' identities and trust each other. In (k, n)-threshold cryptography, the public key of the DCA is known to all the users, while the secret key is divided into many secrete shares among n servers. When a user wants its message signed, the message is sent to all the DCA servers. Each server signs it with its secret share to generate a partial signature. With at least k partial signatures, a valid signature can be constructed that can be verified with the public key. Because these secret shares have no explicit relationship, and can be refreshed periodically without changing the public key, a malicious node has to gather at least k secret shares within some time interval to compromise the system, which makes the system very secure. Besides, the system can tolerate the loss of n - k nodes, it seems ideal for a MANET where nodes may leave or power down unexpectedly [22][23].

5.2. Vulnerability of threshold cryptography-based DCA

In Certificate Authority, the IP address of the node is regarded as the identity of the node. However, in the presence of an autoconfiguration scheme, the identity is generated dynamically. With Sybil attacks from a malicious node, threshold cryptography-based DCA will be compromised, as illustrated below:

Suppose node M is a malicious node in the MANET. Because the IP addresses are assigned with autoconfiguration, node M may request or generate multiple identities in advance. During the formation of a DCA server group based on (k, n)-threshold cryptography, it uses k identities to join the DCA server group with other good nodes, such as nodes A, B, and C. Thus, it has enough secret shares to generate a valid signature for any kind of its own messages.

5.3. Multiple-key cryptography-based DCA

To co-exist with an autoconfiguration scheme, we proposed multiple-key cryptography-based DCA scheme (MC-DCA) in [24]. The idea of multiple-key cryptography was first presented in [25], which is a variation of public-key cryptography. In traditional public-key cryptography, there are only two keys, one of which is the public key and the other is the private key. The message encrypted or signed with one key can be decrypted/verified with the other. In multiple-key cryptography, there are multiple keys. We can choose any subset of the keys to be public keys and the complementary subset will be private keys. The message encrypted/signed with one subset of keys needs to be decrypted/verified with all the keys in the other subset.

MC-DCA scheme is based on a distributed algorithm to generate secret shares and the public key [26]. Suppose there are n servers in the DCA server group. Firstly, all the servers agree on three parameters: two large prime numbers p and q such that q divides p-1, and g that is a generator of Gq (Gq is the unique subgroup of Z*p of order q). These three parameters are a part of the public key and should be known to all the other nodes. Each server (say server i) chooses its secret share xi, and computes the public part of hi = gxi. The sum of xi is the private key, while the product of hi is the public key. The following steps in [26] are used for threshold cryptography and unrelated to MC-DCA scheme.

The protocol of MC-DCA works as follows:

i. When a client needs DCA service, it broadcasts an INVITE message throughout the MANET to initiate the invitation procedure;

ii. On receipt of INVITE message, each node decides if it wants to participate in DCA service. If it wants to join, it broadcasts a PARTCP message with its public key;

iii. All the server nodes agree on the parameters p, q, and g;

iv. Each server node chooses its secret share independently and calculates the corresponding public part. One server node collects all the public part and announces the public key;

v. The client sends its IP address, public key, and other related information encrypted with server node own public key in a REQUEST message;

vi. The server calculates its partial signature for the message, and signs the partial signature with its private key;

vii. The client verifies the signature with the public key after combining all the partial signatures.

If a server node leaves the MANET abruptly, the client can get all the partial signatures. In this case, new nodes would be invited to join the DCA server group. They will choose different secret shares and result in a different public key. Thus, we need to associate a version number with the public key. The client needs to store the all the public keys. We can also require that the certificate be renewed once the version number is increased by a threshold value to remove old public keys.

6. Secure autoconfiguration and public-key distribution

Section 4 and 5 addressed secure autoconfiguration and distributed certificate authority for MANETs separately. This section intends to combine both secure autoconfiguration and public-key distribution when a new node joins the network, which may provide a bootstrapping procedures to build a distributed certificate authority.

6.1. Related work

A self-authentication scheme was proposed in [27] that is an application of Cryptographically Generated Address [28]. A new node generates its public/private key pair randomly and simply uses the hash value of its public key as its IP address. To avoid address conflict, a DAD procedure as described in Section 2 is used. This method is simple and elegant. To verify the ownership of the public key, the other node just needs to perform a hash function its public key. However, there are some problems with this scheme: (1) Only one pair of private/public keys are supported; (2) With the autoconfiguration, the IP address may need to change, which leads to the change of public/private key pair; (3) In case when a MANET is connected to the Internet, the private address of the node may need to be changed with NAT. Thus, there is no relationship with the public key and the public IP address.

The challenge-response scheme [29] is based on the buddy system in [6]. The difference is in the security mechanism in the former: the new node broadcasts its MAC address and public key to its one-hop neighbors for authentication. Because only one-hop neighbors get the public key, other nodes are still susceptible to the "man-in-the-middle" attack. Another issue is that the allocator itself could be a malicious node and allocate a non-disjoint address pool to the new node.

The trust model in [30] is based on MANETconf [7]. It assumes that the number of malicious nodes is small in the network. Each node maintains a trust value for its neighbors. Only the node whose trust value is greater than or equal to a threshold value is considered to be trustworthy. During autoconfiguration, the new node chooses a trustworthy node as an allocator. The allocator also ignores the veto messages from non-trustworthy nodes. However, this scheme is vulnerable to Sybil attacks.

Another trust model proposed in [31] is based on the buddy system [6] and threshold cryptography-based DCA [22]. It assumes that there is already a DCA in the MANET, so the messages can be authenticated. But as we already pointed out in [24], threshold cryptography-based DCA is vulnerable to Sybil attacks.

6.2. Secure autoconfiguration and public-key distribution

Because the identity of the mobile node is generated with the autoconfiguration, it is desirable that the new node's public key is distributed throughout the network at the same time to avoid the "man-in-the-middle" attack. Thus, we combined both in SA-PKD scheme proposed in [32].

The procedures work as follows:

i. The new node, node N, generates its own public/ private key pair (PbN/PrN) and a random number RN;

ii. The new node applies a hash function on the random number RN and gets its address AddrN = Hash(RN);

iii. Node N performs DAD for a few times with a temporary address, as described in [3]. In the DAD message, it puts the hash value of its IP address Hash(AddrN) and the address signed with its private key SignN(AddrN);

iv. On receipt of the DAD message, each node performs the same hash function on its own address. If the result is the same as the hash value in the DAD message, there may be a potential address conflict, so it sends back a NACK message to veto it;

v. If node N does not receive any NACK message, it broadcasts a CMT message to commit the autoconfiguration, in which it puts its public key and uses the address it chooses. On receipt of the CMT message, each node can verify the hash value of the address and the association between the IP address and the public key.

During the broadcast of DAD message, each node gets a copy of the hash value of the address and signature, which can only be verified with the parameters in the subsequent CMT message. If a malicious node uses a random string to replace the parameters in the DAD message, the receiver cannot recover the address. If a malicious node chooses another pair of public/private keys and a random address to replace both parameters in DAD message and CMT message, the receiver will get two associations, including the original one from the new node and a new association of another public key to a different address. In either case, the receiver will get the correct association of PbN to AddrN.

6.3. Simulation

We ran simulation of SA-PKD on ns-2 (version 2.34) with 50 nodes. The random waypoint mobility model is used, in which the nodes are constantly moving within the simulation area. The maximum speed is 20 m/s, and the minimum speed is 5 m/s. The pause time is 0 second. Once the simulation starts, each node joins the MANET every 10 seconds. It broadcasts DAD message and CMT message every 3.0 seconds for 3 times.

We used the MD5 algorithm in [33] for the hash function and a simplified RSA algorithm for signing/verification. We implemented application-level broadcasting, thus there is no preference for the routing protocol.

Except for the first three nodes, we chose malicious nodes randomly. The percentage of malicious nodes is 2%, 4%, 8%, and 10% for different simulations. We let each node print some debug information such as its IP address and public key, and the associations of IP addresses to public keys it received from other nodes. According to the simulation results, all the members can get the associations of new nodes correctly.

7. Conclusion

In this chapter, we gave a comprehensive review of the innovative solutions proposed by the network research community to solve the problems associated with the autoconfiguration. IP address assignment is so important for a node to participate in unicast communications, it is worth our research effort. Unlike a hardwired network, there is no fixed infrastructure in the MANET. Due to node mobility and instability, the network topology keeps changing. In an open system, all kinds of mobile nodes ranging from powerful laptops to energy-efficient sensor nodes may join as long as they confirm to the wireless communication standards. Subsequently, the design of the protocols and algorithms is more complicated.

Autoconfiguration brings a lot of related issues that are trivial to solve or even unseen in a hardwired network. For example, a mobile node may need to change its IP address during the communication, which will break routing protocols and interrupt on-going communications. Security on autoconfiguration and related issues is another important factor to the successful application of the MANET. In Sybil attacks, a malicious node can forge many non-existent fake identities to appear as a group, which can knock out the seemingly robust threshold cryptography-based DCA. All these issues are challenging, and we tried to provide a satisfactory solution that is supported by theoretical analysis and simulation results. They are summarized below:

i. An efficient autoconfiguration solution was proposed for a large-scale MANET;

ii. An IP address handoff scheme aims to reduce the communication overhead caused by address change;

iii. A secure autoconfiguration can withstand several common forms of attacks;

iv. A multiple-key cryptography-based DCA may replace threshold cryptography-based DCA;

v. SA-MKD scheme combines secure autoconfiguration and public-key distribution when a new node joins the MANET.

There is still much work awaiting us. We will continue to investigate more attack patterns on autoconfiguration scheme, study the application of MC-DCA to a large scale network, and explore the possibility of combination of MC-DCA and SA-MKD.

Author details

Hongbo Zhou[1] and Matt W. Mutka[2]

1 Dept. of Computer Science, Slippery Rock University, USA

2 Dept. of Computer Science & Engineering, Michigan State University, USA

References

[1] R. Droms, "Dynamic Host Configuration Protocol," Network Working Group RFC 2131, March 1997

[2] S. Thomson and T. Narten, "IPv6 Stateless Address Autoconfiguration," Network Working Group RFC 2462, December 1998

[3] C. Perkins, J. Malinen, R. Wakikawa, E. Belding-Royer, and Y. Sun, "IP Address Autoconfiguration for Ad Hoc Networks," draft-ietf-manet-autoconf-01.txt, November 2001

[4] N. Vaidya, "Weak Duplicate Address Detection in Mobile Ad Hoc Networks," In Proceedings of the 3rd ACM International Symposium on Mobile Ad Hoc Networking and Computing (MobiHoc'02), Lausanne, Switzerland, June 2002

[5] A.Misra, S. Das, A.McAuley, and S. K. Das, "Autoconfiguration, Registration, and Mobility Management for Pervasive Computing," IEEE Personal Communication, August 2001, pp 24-31

[6] M. Mohsin and R. Prakash, "IP Address Assignment in a Mobile Ad Hoc Network," In Proceedings of MILCOM 2002, Anaheim, CA, October 2002

[7] S. Nesargi and R. Prakash, "MANETconf: Configuration of Hosts in a Mobile Ad Hoc Network," In Proceedings of the 21st Annual Joint Conference of IEEE Computer and Communication Societies (INFOCOM 2002), New York, NY, June 2002

[8] H. Zhou, L. M. Ni, and M. W. Mutka, "Prophet Address Allocation for Large Scale MANETs," In Proceedings of the 22nd Annual Joint Conference of IEEE Computer and Communication Societies (INFOCOM 2003), San Francisco, CA, April 2003

[9] H. Zhou, L. M. Ni, and M. W. Mutka, "Prophet Address Allocation for Large Scale MANETs," Elsevier Ad Hoc Networks Journal, Vol. 1, Issue 4, pp 423-434, November 2003

[10] ns-2 wiki, http://nsnam.isi.edu/nsnam/index.php/Main_Page

[11] G. Pei and M. Gerla, "Mobility management for hierarchical wireless networks," Mobile Networks and Application (MONET), Vol. 6, No. 4, pp 331-337, August 2001

[12] P. Engelstad and G. Egeland, "Name resolution in on-demand MANETS and external IP networks," draft-engelstad-manet-name-resoltuion-00.txt, February 2003

[13] C. Perkins (editor), "IP mobility support," Network Working Group RFC 2002, October 1996

[14] J.-H. Jeong, H.-W. Cha, J.-S. Park, and H.-J. Kim, "Ad hoc IP address autoconfiguration," draft-jeong-adhoc-ip-addr-autoconf-00.txt, May 2003

[15] H. Zhou, M. W. Mutka, and L. M. Ni, "IP Address Handoff in the MANET," In Proceedings of the 23rd Conference of IEEE Communication Society (INFOCOM 2004), Hong Kong, China, March 2004

[16] The netfilter.org project, http://www.netfilter.org/

[17] J. Couceru, "The sybil attack," In Proceedings of the 1st Workshop on Peer-to-Peer Systems (IPTPS'02), Cambridge, MA, March 2002

[18] H. Zhou, "Secure Prophet Address Allocation for Mobile Ad-hoc Networks," In Proceedings of IFIP International Workshop on Network and System Security (NSS 2008), Shanghai, China, October 2008

[19] H. Zhou, M. W. Mutka, and L. M. Ni, "Secure Prophet Address Allocation for MANETs," Wiley International Journal of Security and Communication Networks, Vol. 3, Issue 1, pp 31-34, January/February, 2010

[20] A. Shamir, "How to share a secret," Communications of ACM, Vol. 22, pp. 612-613, November 1979

[21] Y. Desmedt and Y. Frankel, "Threshold Cryptosystems," Proceedings of Advances in Cryptography (Crypto 89), Lecture Notes in Computer Science, Vol. 435, Springer-Verlag, pp. 307 - 315, 1989

[22] L. Zhou and Z. J. Haas, "Securing Ad Hoc Networks," IEEE Network, Vol. 13, No. 6, pp. 24 - 30, November/December 1999

[23] M. Bechler. H.-J. Hof, D. Kraft, F. Pahlke, and L. Wolf, "A Cluster-based Security Architecture for Ad Hoc Neteworks," In Proceedings of the 23rd Conference IEEE Communication Society (INFOCOM 2004), Hong Kong, China, March 2004

[24] H. Zhou, M. W. Mutka, and L. M. Ni, "Multiple-key Cryptography-based Distributed Certificate Authority in Mobile Ad-hoc Networks," In Proceedings of IEEE Global Telecommunications Conference (GLOBECOM 2005), St. Louis, MO, November 2005

[25] C. Boyd, "Some Applications of Multiple Key Ciphers," In Proceedings of Advances in Cryptography (Eurocrypt'88), Lecture Notes in Computer Science, Springer-Verlag, pp. 455 - 467, 1988

[26] T. P. Pedersen, "A Threshold Cryptosystem without a Trusted Party," In Proceedings of Advances in Cryptography (Eurocrypt'91), Lecture Notes in Computer Science, Vol. 547, Springer-Verlag, pp. 522 - 526, 1991

[27] P. Wang,D. S. Reeves, and P. Ning, "Secure Address Autoconfiguration for Mobile Ad Hoc Networks," In Proceedings of the 2nd Annual International Conference on Mobile and Ubiquitous Systems: Networking and Services (MobiQuitous 2005), pp. 519 - 521, San Diego, CA July 2005

[28] T. Aura, "Cryptographically Generated Address (CGA)," Networking Group RFC 3972, March 2005

[29] A. Cavalli and J.-M. Orset, "Secure Hosts Authentication in Mobile Ad Hoc Networks," In Proceedings of the 24th International Conference on Distributed Computing Systems Workshops (ICDCSW 2004), Tokyo, Japan, March 2004

[30] S. Hu and C. J. Mitchell, "Improving IP Address Autoconfiguration Security in MANETs Using Trust Modeling," In Proceedings of 1st International Conference on Mobile Ad-hoc and Sensor Networks (MSN 2005), Wuhan, China, December 2005

[31] F. Buiati, R. Puttini, and R. D. Sousa, "A Secure Autoconfiguration Protocol for MANET Nodes," In Proceedings of the 3rd International Conference on Ad-hoc Networks and Wireless (ADHOC-NOW 2004), Vancouver, Canada, July 2004

[32] H. Zhou, M. W. Mutka, and L. M. Ni, "Secure Autoconfiguration and Public-key Distribution for MANETs," In Proceedings of 6th IEEE International Conference on Mobile Ad-hoc and Sensor Systems (IEEE MASS 2009), Macau SAR, China, October, 2009

[33] R. Rivest, "The MD5 Message-Digest Algorithm," Network Working Group RFC 1321, April 1992

Chapter 7

Privacy-Preserving Information Gathering Using VANET

T. W. Chim, S. M. Yiu, Lucas C. K. Hui and
Victor O. K. Li

Additional information is available at the end of the chapter

http://dx.doi.org/10.5772/53537

1. Introduction

For a driver, some real-time information (e.g. traffic condition along the road and the availability of parking spaces at certain areas) on his way to his destination may be useful. Nowadays, drivers could mainly rely on radio broadcasting. However, the traffic news on radio may not mention anything about the area you are driving into. A more effective way for providing this real-time information would be desirable.

In recent years, a special kind of ad hoc network called Vehicular Ad hoc NETwork (VANET) becomes increasingly popular. It has also become one of the critical components of an Intelligent Transportation Systems (ITS). In a typical VANET, each vehicle is assumed to have an on-board unit (OBU) and there are road-side units (RSU) installed along the roads. A trusted authority (TA) and maybe some other application servers are installed in the backend. The OBUs and RSUs communicate using the Dedicated Short Range Communications (DSRC) protocol [1] over the wireless channel while the RSUs, TA, and the application servers communicate using a secure fixed network (e.g. the Internet). The basic application of a VANET is to allow arbitrary vehicles to broadcast safety messages (e.g. about vehicle speed, turning direction, road condition, traffic accident information) to other nearby vehicles (denoted as vehicle-vehicle or V2V communications) and to RSU (denoted as vehicle-infrastructure or V2I communications) regularly such that other vehicles may adjust their travelling routes and RSUs may inform the traffic control center to adjust traffic lights for avoiding possible traffic congestion. As such, a VANET can also be interpreted as a sensor network because the traffic control center or some other central servers can collect lots of useful information about road conditions from vehicles. It is natural to investigate how to utilize the collected real-time road conditions to provide useful applications. It is natural to consider whether VANET can provide an effective platform for drivers to utilize real-time information collected in RSUs.

In this chapter, we first highlight the most significant security and privacy challenges in VANET protocol design. We then discuss how one should design security protocols for VANETs. For example, we analyze in details the advantages and disadvantages of hardware-based and software-based solutions. Next we propose a VANET-based general information gathering scheme. A driver can issue a query (e.g. road conditions along the roads to his destination) to a nearby RSU, our scheme can then automatically collect the required information from the appropriate RSUs. The gathering process is done in a real-time and distributed manner. Note that the approach of using a centralized server that stores all information collected from RSUs may not work as the information may be changed frequently in a real-time manner and since the VANET is huge, the server will most likely become the bottleneck.

Like other existing VANET applications, there are basic security requirements to be satisfied by such a protocol. They include sender authentication (to ensure that the sender is a valid subscriber), conditional identity privacy preserving (to ensure that a driver's travelling route cannot be traced by any third party except the trusted authority). And there are additional security and privacy requirements to make it more practical. Conditional identity privacy preserving implies that a trusted authority is able to reveal the real identity of a vehicle. If the information gathering scheme is not properly designed, a driver's real identity and query (the information required) can be linked up and analyzed. This is not preferable because we want to ensure that no one in this world (including trusted authority) knows what a driver is querying for. This leads to our privacy preserving problem. Besides, a driver may not want vehicles nearby to know his query by eavesdropping his message. Also when the system sends the result back to him/her, we do not want non-subscribers nearby to enjoy free information gathering service in case it is a paid service. We regard this as a confidentiality problem. Finally, since our information gathering scheme involves the information provided by more than one RSU and RSUs are left unattended at roadsides most of the time, proper and efficient authentication of this information becomes critical. Our scheme addresses this authentication problem as well.

We provide a security analysis and a simulation study to evaluate our scheme. In our simulation, we make use of the maps of New York (city road system) and California (countryside highway system) downloaded from the TIGER database. We find that the processing time is at most 1.6 % and 3.7 %, respectively, of the duration that the vehicle stays in the querying RSU's range in the two cities. Thus there must be sufficient time for the vehicle to finish its query and to verify the returning information.

The rest of this chapter is organized as follows: Challenges of security protocols for VANETs are discussed in Section 2. Hardware and software approaches are then explained and compared in Section 3. The system model and the problem statement are described in Section 4. Some preliminaries on bilinear map are given in Section 5. Our schemes are presented in Section 6. The analysis and evalution of our schemes are given in Sections 7 and 8. Related work is reviewed in Section 9. Finally, Section 10 concludes the chapter.

2. Challenges of security protocols for VANETs

General security vulnerabilities and challenges for VANETs have been discussed in works like [2] and [3]. On the other hand, we focus on the challenges for designing security protocols in VANETs.

1. Dynamic, linear and real-time topology

   Moving vehicles are major components of VANETs. They are moving at high speed most of the time and this makes a VANET topology change rapidly and subject to frequent fragmentation. A vehicle which connects part of the VANET at a certain moment may no longer act as a connector in the next moment. Also, unlike MANET, nodes move in random direction, VANET vehicles move in a constrained manner. A vehicle must move along roads and change its direction only at junctions. Vehicles on a road tend to alight in a straight line. Security protocols for VANETs should not assume any fixed node infrastructure such as trees [4] [5]. Instead, dynamic topology should be properly handled. Furthermore, a VANET topology could be affected by drivers' reaction to messages. For example, a driver may change its route after receiving a message about congestion from another vehicle. Therefore, all tasks including those for security purpose should be performed in real-time. Centralized pre-processing is not possible.

2. Large scale and density varying network

   A VANET usually covers the whole region or even the whole city and thus the total number of VANET nodes can be very huge. This means that a centralized security protocol such as [6] may not be a good choice. Instead, operations have to be done in a decentralized and distributed manner. On the other hand, a VANET usually has different network density in different regions. For example, at where there is a traffic jam, the network becomes very dense. On the contrary, in suburban area, the network becomes very sparse. This implies when designing security protocols for VANETs, we cannot have the assumption of low or high network density. Instead, a good protocol should be able to handle both situations.

3. Transmission and computation efficiency

   One of the initial design goals of VANETs is the sharing of critical information (e.g. to inform vehicles about danger ahead of a road) among vehicles. Thus most messages in VANETs are of real-time importance. Therefore, security operations should cause as low overhead to the network as possible. In recent years, researchers start adopting elliptic curve cryptography (ECC) approach [7] to reduce key and ciphertext sizes. Also some other researchers are trying to reduce the computation overhead induced by security operations. For example, authors in [8] and [9] propose an efficient batch signature verification technique.

4. Conditional identity privacy perserving

   Normally, a driver may not want others to know his real identity and then trace his route or driving habit. If this cannot be satisfied, he may not subscribe to any new service including VANET at all. Thus the real identity of any vehicle should be kept anonymous from others and a third party should not be able to reveal a vehicleąęs real identity by analyzing multiple messages sent by it. RSUs are just installed along the roadside and are more vulnerable to attack. In the extreme case, even if all RSUs collude, we want to make sure that the relationship between the real identity of a vehicle and the messages it sent cannot be revealed.

   However, since vehicles are fast moving objects, injuries or even death are usually caused when accidents take place. If an accident is caused by a VANET message, a trusted party such as the police force may need to find out the message sender so as to avoid repeated occurrence of similar accidents. Thus while preserving a vehicleąęs privacy, its real identity should be able to be traced by a trusted party when necessary. Thus we call the identity privacy preserving here conditional.

5. Small network diameter

   Because of the dynamic nature of VANETs, the network diameter, which is defined as the number of hops between the furthest endpoints of the network, can be very small. The network route between two vehicles may be disconnected easily due to moving out-of-range or having obstacle blocking. Thus those secure routing protocols originally designed for fixed Internet or MANETs cannot be directly adopted into VANETs. New protocols such as 'carry and forward' [10] have to be adopted to fit this specific property. Security issues induced by this kind of forwarding strategy are still open problems.

6. Multiple levels of security

   Different kinds of messages can exist in a nowadays VANET. As mentioned earlier, most messages are critical and are about conditions on the road. However, there can be others such as advertisements [11]. Thus a good security protocol should provide multiple security levels such as what is proposed by [12]. A critical message should have better protection than an advertisement message.

7. Energy efficiency considerations

   Unlike mobile ad hoc networks (MANETs) and wireless sensor networks (WSNs) where nodes are assumed to run on self batteries, energy is no longer a challenge in VANETs. It is because OBUs are continuously charged by car batteries while RSUs are continuously charged by fixed power cables. Thus researchers' attentions should be shifted back to security problems themselves rather than paid to energy efficiency directions.

3. Hareware and software approaches

Recent security works for VANETs go for a certificate-less direction (i.e. a sender does not need to send its certificate to the receiver for verification). However by nature, a vehicle's signature must contain a secret that is known by the receiver to facilitate the validation. To accomplish this, recent works focus on two major directions - hardware-based and software-based. In this section, we first explain how hardware-based and software-based solutions work respectively. Then we briefly discuss the advantages and disadvantages of each approach.

Hardware-based solutions usually rely on a tamper-proof device installed on a vehicle [3] [8]. It contains the secret we mentioned earlier and runs on its own battery and own clock. Thus an outsider cannot block its functions by cutting its power supply or by inputting wrong signals. Besides storing secrets, it is also in charge of all security operations including digital signature, encryption and decryption. Further, the device is accessible only by authorized personnel. A driver has to input a password before the device can function properly. A recent example of hardware-based solutions is [8]. Here the tamper-proof device installed on a vehicle has an accessing password preloaded. This password is assigned by a TA and is firmed burned onto the hardware when the device is first registered. Whenever a driver starts the vehicle, he/she has to input into the device the same accessing password in order to enable it for further operations. This is how the authentication of driver by the device is done. Besides the accessing password, the tamper-proof device also stores all system public

http://dx.doi.org/10.5772/53537

parameters together with the master keys s1 and s2 of the TA. These master keys are assumed to be known by only vehicles and the TA but not RSUs. They are used to form the signing keys for constructing signatures. TA's master keys also facilitates RSUs to verify vehicles' signatures even though they do not have any knowledge about the master keys. (It is based on pairing operations and interested readers please refer to [8] for details.) Note that the master keys can only be used for security operations inside the tamper-proof device. No outsiders, including the driver, know about their values.

Software-based solutions, on the contrary, do not rely on any tamper-proof device. What a vehicle has is an ordinary computer device. In other words, no secret or parameter can be preloaded securely onto a vehicle. However, it still requires them for security operations like digital signatures. Software-based solutions obtain these information through a secure initial handshaking. A recent software-based solution is presented in [9]. A conventional public key infrastructure is assumed to exist for initial secure message exchange. Whenever a driver starts the vehicle, he/she has to input into the computer device an accessing password. This password is pre-assigned by the TA and is assumed to be given to the driver earlier (e.g. via paper documents during car first registration). The password input by the driver is then encrypted using the TA's conventional public key which is assumed to be known by everyone. The TA decrypts the password using its conventional private key and checks whether it matches with its records. If yes, it encrypts its master keys using the vehicle's conventional public key. Upon receiving the encrypted message, the vehicle can obtain those master keys by decrypting the encrypted block using its conventional private key. A vehicle's conventional public and private key pairs are also assumed to be assigned by the TA at earlier stage (e.g. during car first registration). Thus the computer device can perform security operations as what a tamper-proof device does. Based on the above brief descriptions, we can see that both hardware-based and software-based solutions can resolve the challenges we mentioned in Section 2:

1. They do not have any assumptions on the network nature and so they can fit the dynamic, linear, density-varying and small-diameter VANET topology well.
2. Except initial handshaking, all cryptographic operations are done in real-time and so they are suitable for real-time-changing VANET environments.
3. Both of them achieve transmission and computation efficiency. First, they adopt ECC which possess the property of short key. Second, efficient signature batch verification routines were proposed (please refer to [8] and [9] for details). Third, unlike traditional public key infrastructure, a vehicle does not need to send its certificate to others for signature verification purpose. Finally, unlike mobile phone network which is mainly for unicast, Dedicated Short Range Communications (DSRC) [1], which are short to medium range wireless communications channels specifically designed for automative use, can facilitate efficient broadcast.
4. A vehicle only attaches a pseudo identity in its messages. Its real identity can only be traced by TA using a tracing routine (please refer to [8] and [9] for details). Thus conditional identity privacy can be preserved.
5. Regarding multiple levels of security, there is already a representative work [12]. Thus, the extension is not difficult and is practical actually.

Next let us compare the two approaches from a number of aspects.

1. Authentications of drivers

   For hardware-based solutions, to authenticate a driver, a tamper-proof device only needs to check the accessing password input by the driver locally. If it does not match the burned one, it simply disables all its functions. However, for software-based solutions, the ordinary computer device does not know whether the accessing password input by the driver is correct or not because it has no secret pre-stored. Hence, it needs to securely transmit it via network to the TA for further verification.

2. System parameters preloading

   For hardware-based solutions, the system-wide TAąęs master keys are preloaded into a tamper-proof device. Thus, the TA does not need to send them to the device anymore after an initial hardware burning. However, for software-based solutions, no secret or parameter is stored in the ordinary computer device. Thus every time a driver starts the vehicle and after accessing password checking, the TA has to send them again to the device. Extra transmission overhead is needed in each session.

3. Replication of device contents

   One basic assumption of tamper-proof device or smart card technology is that the contents inside the device or smart card cannot be improperly extracted or replicated easily. Thus secrets and parameters stored on them can be said to be fully protected. However, the case is not the same for ordinary computer device in software-based solutions. As everyone knows, the contents of a hard-disk can be cloned or replicated easily. This is why in the software-based solutions mentioned earlier, system secrets cannot be stored in a computer device. Instead they have to be transmitted from the TA every time it starts up.

4. Updates of system parameters

   For hardware-based solutions, the system-wide TAąęs master keys and other public parameters cannot be updated easily. Once an update is needed, a driver has to physically bring the device to the TA for an update. However, for software-based solutions, all secrets and public parameters are transmitted from the TA in real time when the vehicle starts up. Thus updates can be easily done. All the TA needs to do is to send the new set of secrets and parameters to the device.

5. Setting of new secrets after compromise

   For hardware-based solutions, the same set of system-wide TAąęs master keys and public parameters are preloaded into all tamper-proof devices. Once one of the devices is cracked by an attacker, the whole VANET system will be compromised unless a physical hardware update by the TA is done. However, software-based solutions do not have this problem. When one of the devices is found to be cracked, the TA can invoke an update of master keys and parameters by simple and secure network transmissions.

6. Modification of protocols

   For hardware-based solutions, all security operations are carried out by the processor of a tamper-proof device. Thus the same set of security operations are preloaded into all devices. If, in the mean time, the TA wants to introduce a new security operation (e.g. to enhance the security level of the VANET system), it needs to ask all drivers to bring the devices to it for a hardware update. However, software-based solutions do not have this problem. When the TA wants to introduce a new security operation, it only needs to enable all computer devices to securely download a new software (like how we update our computer operating systems nowadays).

http://dx.doi.org/10.5772/53537

Features	Hardware-based	Software-based
Authentications of drivers	Local tamper-proof device only	Transmission to TA needed
System-wide secrets and public parameters preloaded	Yes	No
Replication of device contents	Very difficult	Relatively easier
Updates of system-wide secrets and public parameters	Physical hardware update at TA	Simple secure download
Setting of new secrets after compromise	Physical hardware update at TA	Simple secure download
Modification of protocols	Physical hardware update at TA	Simple secure download
Complexity of security operations	Need to be simple	Relatively more complicated

Table 1. Hareware-based vs. Software-based Solutions

7. Complexity of security operations

   For hardware-based solutions, all security operations are carried out by the processor of a tamper-proof device. Unluckily, the computation power of the processor is quite limited. Up to our knowledge, not all smart cards in the market today are powerful enough to perform pairing operations. Thus, security operations adopted have to be as simple as possible. However, software-based solutions do not have this limitation. Even the poorest CPU today can handle complicated operations like pairing in reasonable time.

We summarize the comparisons between hardware-based, hybrid and software-based solutions in Table 1. We can see that the items are around efficiency, flexibility and security. In short, recall that a VANET is of large scale, hardware-based solutions are more efficient than software-based ones since they can reduce the transmission overhead between devices and TA during initial handshaking. However, there may still be problems when updating of system parameters, secrets or cryptographic protocols is required since if all drivers go to the TA for parameters, hardware or software updates, a bottleneck will appear. Our proposed scheme is hardware-based but we also provide suggestions about how to update system parameters, secrets or cryptographic protocols efficiently.

4. System model and assumptions

Besides the assumptions made in other VANET applications such as TA being trusted and real identity of any vehicle being known by TA and itself but not by others, we further assume the followings:

1) There exists a conventional identity-based public key infrastructure (PKI). The public key of the TA is the same as its real identity $TRID$ and is known by *everyone*. Also any RSU R_i broadcasts its public key which is the same as its real identity $RRID_i$ with hello messages periodically to vehicles that are travelling within the RSU-Vehicle Communications (RVC) range of it. The validity of $RRID_i$ can be ensured using a certificate issued by the TA.

2) Each RSU has a local database storing road information in its range (e.g. GPS locations of boundaries, names of buildings and streets, etc.). This facilitates an RSU to answer queries that are about fixed facilities in its range.

3) Each vehicle has a tamper-proof device and a conventional computer device with GPS receiver. The tamper-proof device is responsible for generating pseudo identities and signatures on messages (details will be given later in the next section) and is assumed to have its own clock for generating correct time stamps and be able to run on its own battery [3]. The conventional computer device is responsible for all other calculations and can receive GPS signals.

4) We assume that there is a reasonably large number of information gathering queries issued to RSUs. Otherwise, if there is only one query, the sender can be linked up with the query easily.

5. Preliminaries

Our security scheme is *pairing-based* and defined on two cyclic groups with a mapping called *bilinear map* [7]. In this section, we briefly introduce what a bilinear map is.

Let $\mathbb{G}$ be a cyclic additive group and $\mathbb{G}_{\mathbb{T}}$ be a cyclic multiplicative group. Both groups $\mathbb{G}$ and $\mathbb{G}_{\mathbb{T}}$ have the same prime order q. The mapping $\hat{e} : \mathbb{G} \times \mathbb{G} \rightarrow \mathbb{G}_{\mathbb{T}}$ is called a bilinear map if it satisfies the following properties:

1. Bilinear: $\forall P, Q, R \in \mathbb{G}$ and $\forall a, b \in \mathbb{Z}$, $\hat{e}(Q, P + R) = \hat{e}(P + R, Q) = \hat{e}(P, Q) \cdot \hat{e}(R, Q)$. Also $\hat{e}(aP, bP) = \hat{e}(P, bP)^a = \hat{e}(aP, P)^b = \hat{e}(P, P)^{ab}$.
2. Non-degenerate: There exists $P, Q \in \mathbb{G}$ such that $\hat{e}(P, Q) \neq 1_{\mathbb{G}_{\mathbb{T}}}$.
3. Computable: There exists an efficient algorithm to compute $\hat{e}(P, Q)$ for any $P, Q \in \mathbb{G}$.

The bilinear map $\hat{e}$ can be constructed using pairings on elliptic curves. Each operation for computing $\hat{e}(P, Q)$ is referred as a pairing operation. Pairing operation is the most expensive operation in this kind of cryptographic schemes. The fewer the number of pairing operations, the more efficient the scheme is. The groups $\mathbb{G}$ and $\mathbb{G}_{\mathbb{T}}$ are called bilinear groups. The security of our schemes relies on the fact that the discrete logarithm problem (DLP) on bilinear groups is computationally hard, i.e., given the point $Q = aP$, there exists no efficient algorithm to obtain a by given P and Q. The implication is that we can transfer Q in an open wireless channel without worrying that a (usually some secret) can be known by any attackers.

6. Our scheme

This section presents our Privacy-preserving Information Gathering scheme. We first summarize our scheme into some basic steps (see Figure 1):

1) TA sets up parameters and generates anonymous credentials.

2) Vehicle V_i requests for a credential from RSU R_j.

3) RSU R_j verifies V_i's identity and sends it a credential.

4) After a random delay or after travelling for a random distance, V_i sends out its request to RSU R_k.

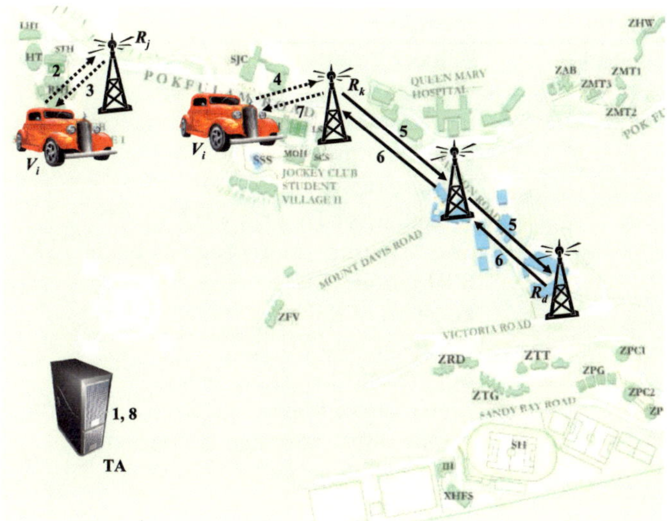

Figure 1. Basic Steps in Our Scheme

5) RSU R_k forwards the request to its neighbors. This process repeats until the request reaches the RSU covering the furthest point of interest with respect to V_i's current location.

6) RSU R_d constructs the information reply message and sends it along the reverse path. Each hop whose range overlaps with the region of interest attaches the corresponding hop information (with signature).

7) RSU R_k forwards the reply message to V_i which then verifies the messages from all RSUs along the route in a batch.

8) Based on V_i's pseudo identity received from RSU R_j, TA reveals V_i's real identity for billing purpose.

Next we explain our scheme in details. The notations used in this chapter are summarized in Table 2.

6.1. Setup

During system startup, the following steps will be carried out by TA:

1. It chooses groups $\mathbb{G}$ (with g as the generator) and $\mathbb{G}_{\mathbb{T}}$ that satisfy bilinear map properties.
2. It randomly picks $s \in \mathbb{Z}_q$ as the master secret (preloaded into all vehicles' tamper-proof devices).
3. It computes $g_{pub} = g^s$ as a public parameter. Note that given $g_{pub} = g^s$, there exists no efficient algorithm to obtain s based on the fact that the discrete logarithm problem (DLP) on bilinear groups is computationally hard.
4. It assigns itself a secret key TSK and an identity $TRID = g^{TSK}$ which is assumed to be known by everyone in the system.

Symbol	Meaning
$\mathbb{G}$ and $\mathbb{G_T}$	Bilinear groups
g	Generator of $\mathbb{G}$
s	System master secret
$g_{pub} = g^s$	Public parameter
$TRID$	Identity of TA
TSK	Secret key of TA s.t. $TRID = g^{TSK}$
$TSIG_{TSK}(M)$	TA's signature on message M using TSK
R_i	RSU number i
RL_i	Location of RSU R_i
RC_i	Certificate of RSU R_i
$RRID_i$	Identity of RSU R_i
RSK_i	Secret key of RSU R_i s.t. $RRID_i = g^{RSK_i}$
C_T	Anonymous credential for period T
V_i	Vehicle number i
VC_i	Certificate of vehicle V_i
CPK_i	Conventional public key of vehicle V_i
CSK_i	Conventional private key of vehicle V_i
$VRID_i$	Real identity of vehicle V_i
$VPWD_i$	Hardware activation password on V_i
$VPID_i$	Pseudo identity of vehicle V_i
VSK_i	Signing key of vehicle V_i
$S\_ENC_x(M)$	Symmetrical encryption of M using key x
$AS\_ENC_x(M)$	Asymmetrical encryption of M using key x
$SIG_x(M)$	Signature on message M using key x
$H(M)$	MapToPoint hash value [13] on message M
$h(M)$	One-way hash value of message M

Table 2. Notations used in this chapter

5. It assigns each RSU R_i locating at RL_i a secret key RSK_i, an identity $RRID_i = g^{RSK_i}$ and generates its certificate as $RC_i =< RRID_i, RL_i, TSIG_{TSK}(RRID_i||RL_i) >$ where $TSIG_{TSK}(RRID_i||RL_i) = H(RRID_i||RL_i)^{TSK}$ is TA's signature on the concatenation of $RRID_i$ and RL_i. Here $H(.)$ is a MapToPoint hash function.

6. It assigns each vehicle V_i a real identity $VRID_i = g^x$ where x is a random number and can be thrown away after generating $VRID_i$, and the hardware activation password $VPWD_i$. TA preloads them into the tamper-proof device of V_i.

7. It assigns each vehicle V_i a pair of conventional public key $VCPK_i$ and private key $VCSK_i$ under any public key infrastructure. $VCSK_i$ is preloaded into the tamper-proof device of V_i while $VCPK_i$ is stored into TA's local database. This conventional public and private keys are for updating the master secret s when there is a need (e.g. when any vehicle is proved to be compromised and the master secret is leaked to attackers). During such an update, TA can encrypt and send the new master secret to each uncompromised vehicle V_i using the corresponding $VCPK_i$. In this way, only the uncompromised vehicles can decrypt and obtain the new master secret.

Throughout this chapter, let us use the notations $AS\_ENC_x(M)$ and $S\_ENC_x(M)$ to denote encrypting message M using the key x based on any asymmetric and symmetric encryption algorithms, respectively.

6.2. Generation of anonymous credentials by TA

In our scheme, a credential will expire after a predefined period of time. Thus even if a subscriber leaks its credential to a non-subscriber or even to an attacker, the impact to the system is limited. Assume that the current time is T. TA computes the credential for the current period as $C_T =< \mathbf{CRD}, T, TSIG_{TSK}(\mathbf{CRD}||T) >$, where $TSIG_{TSK}(\mathbf{CRD}||T) = H(\mathbf{CRD}||T)^{TSK}$, and sends it to all RSUs securely via a fixed infrastructure. We can see that the credential carries no information about any user and that is why we call it "anonymous".

6.3. Activation of tamper-proof device on vehicle V_i

When the vehicle V_i starts, the driver enters the real identity $VRID_i$ and password $VPWD_i$ (assigned by TA in Section 6.1) into the tamper-proof device to activate it. Here only simple hardware checking is involved. The tamper-proof device continues with its pseudo identity generation and message signing tasks only if both the real identity and the password are correct. That means V_i cannot use the service if it is being stolen.

6.4. Vehicle V_i requesting for anonymous credential at RSU R_j

To request for an anonymous credential, V_i's tamper-proof device performs the following steps:

1. It generates a pseudo identity $VPID_i = (VPID_{i1}, VPID_{i2}) = (g^r, VRID_i \oplus H(g^r_{pub}))$ where r is a per-session random nonce.
2. It composes the credential request message $M_i = \{\mathbf{CRD\_REQ}\}$.
3. It picks a random number $rand$ and encrypts it using R_j's identity as $AS\_ENC_{RRID_j}(rand)$. This random number becomes a shared secret between itself and RSU R_j. R_j will use it to encrypt the credential at a later stage.
4. It generates the signing key $VSK_i = (VSK_{i1}, VSK_{i2}) = (VPID^s_{i1}, HP^s_i)$ where $HP_i = H(VPID_{i1}||VPID_{i2})$.
5. It generates the signature σ_i on M_i and T_i (T_i is the current timestamp given by the tamper-proof device) as $VSK_{i1} \times VSK_{i2}^{h(M_i||T_i)}$ where $h(.)$ is a one-way hash function such as SHA-1.
6. It sends $< AS\_ENC_{RRID_j}(rand), VPID_i, M_i, T_i, \sigma_i >$ to RSU R_j nearby.

The RSU R_j then performs the following steps:

1. It checks the timestamps in the messages. For any message, if the difference between the attached timestamp and the current time is larger than a threshold (which is a system parameter), the message is ignored. This can help reduce the impact of reply attack.

2. It verifies V_i's signature by checking whether $\hat{e}(\sigma_i, g) = \hat{e}(VPID_{i1} \times HP_i^{h(M_i||T_i)}, g_{pub})$.
   Proof of correctness:
   L.H.S.
   $= \hat{e}(\sigma_i, g)$
   $= \hat{e}(VSK_{i1} \times VSK_{i2}^{h(M_i||T_i)}, g)$
   $= \hat{e}(VPID_{i1}^s \times HP_i^{sh(M_i||T_i)}, g)$
   $= \hat{e}(VPID_{i1} \times HP_i^{h(M_i||T_i)}, g^s)$
   $= \hat{e}(VPID_{i1} \times HP_i^{h(M_i||T_i)}, g_{pub})$
   = R.H.S. □
3. If it receives requests from more than one vehicle at the same time (say request messages M_{first}, ..., M_{last}, signatures σ_{first}, ..., σ_{last} from vehicles V_{first}, ..., V_{last} respectively), it verifies them in a batch by checking whether $\hat{e}(\prod_{i=first}^{last} \sigma_i, g) = \hat{e}(\prod_{i=first}^{last} VPID_{i1} \times HP_i^{h(M_i||T_i)}, g_{pub})$.
   Proof of correctness:
   L.H.S.
   $= \hat{e}(\prod_{i=first}^{last} \sigma_i, g)$
   $= \hat{e}(\prod_{i=first}^{last} VSK_{i1} \times VSK_{i2}^{h(M_i||T_i)}, g)$
   $= \hat{e}(\prod_{i=first}^{last} VPID_{i1}^s \times HP_i^{sh(M_i||T_i)}, g)$
   $= \hat{e}(\prod_{i=first}^{last} VPID_{i1} \times HP_i^{h(M_i||T_i)}, g^s)$
   $= \hat{e}(\prod_{i=first}^{last} VPID_{i1} \times HP_i^{h(M_i||T_i)}, g_{pub})$
   = R.H.S. □
4. For each vehicle whose signature is valid, R_j encrypts the anonymous credential for the current period C_T using *rand* and sends $S\_ENC_{rand}(C_T)$ back to it.

6.5. Vehicle V_i requesting for information at RSU R_k

Note that if V_i obtains the credential C_T from RSU R_j and if it sends out its query to R_j immediately, its real identity and its query can always be linked up once R_j colludes with TA. Thus we propose two approaches to avoid this from happening:

1. V_i sends out its query to R_j only after a random delay. This is because under normal situation, there will be credential requests from other vehicles during that random period and as a result R_j cannot link up which query belongs to which credential request.
2. V_i sends out its query at another RSU (say $R_k \neq R_j$) after travelling for a random distance. Since R_k does not know V_i's credential request (thus pseudo identity), even if it colludes with TA, it cannot link up V_i's real identity and its query.

Now assume that V_i sends its query to RSU R_k. V_i performs the following:

1. It composes the request message $M_i = \{\textbf{SREQ}, LOC_i, Interest_i\}$ where LOC_i represents the current location of V_i and $Interest_i$ contains a set of points of interest (in GPS coordinates) and description of information required (e.g. average vehicle speed, congestion status).

2. It picks two random numbers *rand* and *sn*. *rand* is for R_k to encrypt the result at a later stage and *sn* is used as a session number.
3. It sends $< AS\_ENC_{RRID_k}(rand, sn, C_T, M_i) >$ to R_k and stores *rand* and *sn* locally.

R_k then performs the following steps:

1. It decrypts the message using its private key.
2. It ensures the credential used C_T is not outdated (e.g. the timestamp should be within a pre-defined number of periods before the current time).
3. It verifies TA's signature on C_T by checking whether $\hat{e}(TSIG_{TSK}(\mathbf{CRD}||T), g) = \hat{e}(H(\mathbf{CRD}||T), TRID)$.
   Proof of correctness:
   L.H.S.
   $= \hat{e}(TSIG_{TSK}(\mathbf{CRD}||T), g)$
   $= \hat{e}(H(\mathbf{CRD}||T)^{TSK}, g)$
   $= \hat{e}(H(\mathbf{CRD}||T), g^{TSK})$
   $= \hat{e}(H(\mathbf{CRD}||T), TRID)$
   = R.H.S. □
4. It the signature is valid, it proceeds to the information gathering process.
5. It stores *rand* and *sn* locally for later usage.

6.6. Request and reply propagation

RSU R_k takes up the role of initiating the information gathering process by composing the information request message $M_k = \{\mathbf{INFO\_REQ}, sn, RRID_k, LOC_i, Interest_i\}$. Let FP be the furthest point with respect to LOC_i in $Interest_i$. R_k broadcasts M_k to all neighbors which are closer to FP than itself.

Any receiving RSU first stores sn, $RRID_k$ and $Interest_i$ into its routing table to build up the reverse path so that it can send any reply back to R_k later on. Let FP be the furthest point with respect to LOC_i in $Interest_i$. It then checks whether FP is within its range. If not, it simply re-broadcasts M_k to all neighbors which are closer to FP than itself. Otherwise, it computes the information reply message $M_d = \{\mathbf{INFO\_RPY}, sn, RRID_d, RL_d, RC_d, HopInfo_d, \sigma_d\}$ and sends it back to its previous RSU hop. Here $HopInfo_d$ is the information that is of V_i's interest and $\sigma_d = H(HopInfo_d)^{RSK_d}$ is R_d's signature on $HopInfo_d$.

Each RSU hop along the reverse path R_{im} repeats the steps done by R_d and if any point in $Interest_i$ is within its range, it includes information and signature corresponding to its hop (i.e. $HopInfo_{im}$ and σ_{im}) into the information reply message. Otherwise, it simply forwards the reply message to its previous RSU hop.

Upon receiving a reply, R_k encrypts it using *rand* and forwards it to V_i immediately.

6.7. Verification of RSUs' hop information

Recall that vehicle V_i receives from R_k a set of identities, a set of hop information and a set of signatures, each corresponding to an RSU along the path of propagation. To verify the hop information provided by an RSU, its signature is verified using its identity. In turn, to verify an RSU's real identity, its certificate has to be verified using TA's identity.

Let us first talk about how the RSUs' certificates can be verified in a batch. Without loss of generality, assume the RSUs along the returned route have real identities $RRID_{first}$, ..., $RRID_{last}$, locations RL_{first}, ..., RL_{last} and TA signatures $TSIG_{TSK}(RRID_{first}||RL_{first})$, ..., $TSIG_{TSK}(RRID_{last}||RL_{last})$. Vehicle V_i can then verify the $(last - first + 1)$ signatures in a batch by checking whether $\hat{e}(\prod_{i=first}^{last} TSIG_{TSK}(RRID_i||RL_i), g) = \hat{e}(\prod_{i=first}^{last} H(RRID_i||RL_i), TRID)$

Proof of correctness:
L.H.S.
$= \hat{e}(\prod_{i=first}^{last} TSIG_{TSK}(RRID_i||RL_i), g)$
$= \hat{e}(\prod_{i=first}^{last} H(RRID_i||RL_i)^{TSK}, g)$
$= \hat{e}((\prod_{i=first}^{last} H(RRID_i||RL_i))^{TSK}, g)$
$= \hat{e}(\prod_{i=first}^{last} H(RRID_i||RL_i), g^{TSK})$
$= \hat{e}(\prod_{i=first}^{last} H(RRID_i||RL_i), TRID)$
= R.H.S. □

Further assume these $(last - first + 1)$ RSUs provide the hop information $HopInfo_{first}$, ..., $HopInfo_{last}$ together with signatures $(\sigma_{first}, ..., \sigma_{last}$. Vehicle V_i verifies these signatures in a batch by checking whether $\hat{e}(\prod_{i=first}^{last} \sigma_i, g) = \prod_{i=first}^{last} \hat{e}(H(HopInfo_i), RRID_i)$.

Proof of correctness:
L.H.S.
$= \hat{e}(\prod_{i=first}^{last} \sigma_i, g)$
$= \prod_{i=first}^{last} \hat{e}(\sigma_i, g)$
$= \prod_{i=first}^{last} \hat{e}(H(AvgSpd_i||RoadCond_i)^{RSK_i}, g)$
$= \prod_{i=first}^{last} \hat{e}(H(AvgSpd_i||RoadCond_i), g^{RSK_i})$
$= \prod_{i=first}^{last} \hat{e}(H(AvgSpd_i||RoadCond_i), RRID_i)$
= R.H.S. □

We can see that vehicle V_i needs to perform only 2 pairing operations to verify the certificates of all RSUs. For the message verification, since the signatures are generated by different RSUs, altogether $(last - first + 2)$ pairing operations are needed. Note that the above verification procedures still apply even if the returned route contains only one single hop R_k. In that case, we can simply set $first = last = k$ in the expressions.

6.8. Traceability of vehicle V_i's real identity

With V_i's pseudo identity $VPID_i = (VPID_{i1}, VPID_{i2}) = (g^r, VRID_i \oplus h(g_{pub}^r))$ and the master secret s, TA can retrieve V_i's real identity by computing $VRID_i = VPID_{i2} \oplus h(VPID_{i1}^s)$.

7. Security analysis

We analyze our scheme with respect to the security and privacy requirements mentioned earlier.

1) Conditional identity privacy preserving: The pseudo identity of any vehicle is an ElGamal-type ciphertext, which is secure under the chosen plaintext attacks [14]. Also the random nonce r makes them different in different messages. To trace the real identity, one needs to know the value of s but s is only known by all tamper-proof devices and TA. A tamper-proof device (which can prevent unauthorized parties from modifying its logic or reading its stored data) is not supposed to carry out such a traceability function. On the other hand, Section 6.8 shows that TA is able to trace a vehicle's real identity. Thus no one except TA can trace the real identity of a particular vehicle and conditional identity privacy is achieved.

2) Privacy preserving and unlinkability: After vehicle V_i obtains an anonymous credential, it presents it to the same RSU after a random delay or to a different RSU for service as discussed earlier. In any case, that RSU does not know V_i's pseudo identity and identity verification is based on an anonymous credential, it cannot link up V_i's query with its identity even if it colludes with TA. Thus unlinkability is guaranteed.

3) Confidentiality: When vehicle V_i requests for a credential from RSU R_j, it first picks a random number $rand$ and securely sends it to R_j. R_j in return encrypts the credential using $rand$. Thus neighboring vehicles cannot illegally receive the credential by eavesdropping messages from the air. Similarly, when vehicle V_i requests for information gathering service from RSU R_k, it picks another random number and R_k in return encrypts the result using that random number. Thus no other vehicles can eavesdrop the result even if they are interested in similar information. For the query, V_i encrypts it using RSU's identity and so it is kept confidential from others.

4) Message authentication: TA's signature on message M is defined as $H(M)^{TSK}$. Since TSK is only known by TA, no others can forge the signature.

Similarly, RSU R_j's signature on message M is defined as $H(M)^{RSK_j}$. Again since RSK_j is only known by R_j, no others can forge the signature.

Regarding vehicle V_i's signature, it is composed of VSK_{i1} and VSK_{i2}. VSK_{i1} is defined as g^{rs}. Due to the difficulty of solving the discrete logarithm problem, there is no way for attackers to obtain s and thus no one other than the tamper-proof device can compose VSK_{i1}. VSK_{i2}, on the other hand, is defined as HP_i^s. Again, since no one other than tamper-proof devices knows s, VSK_{i2} cannot be forged as well.

8. Simulation results

In this section, we evaluate our scheme in terms of processing delay using a network simulation program. Through simulation, we show that the processing delay caused by our cryptographic functions is minimal.

8.1. Simulation models

In our simulation, we made use of two maps downloaded from the TIGER database [15] - one is New York and the other is California. New York represents a city road system (see Fig. 2 for the Google Map [16]) in which most roads have speed limit of 50 km/h. California, on the other hand, represents a countryside road system (see Fig. 3 for the Google Map [16]) in which some highways have speed limit up to 120 km/h. RSUs are randomly placed onto each road. With the consideration of speeding behavior, we assume New York has average vehicle speed readings from 0 km/h (road blocking situation) to 70 km/h (speeding situation) while California has average vehicle speed readings from 0 km/h (highway blocking situation) to 140 km/h (speeding situation).

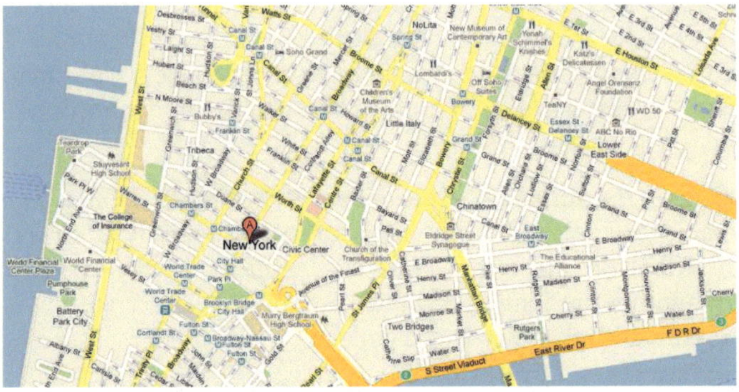

Figure 2. City Road System in New York

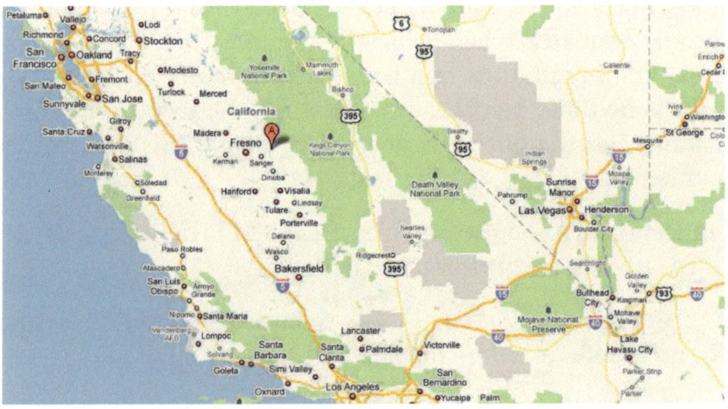

Figure 3. Countryside Highway System in California

The settings and parameters of our simulation are adopted from [8] and [9]. Interested readers may refer to them for details. We fix the size of our newly-introduced components as follows: 5 bytes for control messages like **CRD_REQ**, 20 bytes for each representation of GPS location, 255 bytes for timestamp and 10 bytes for random number.

We define 16 geographical distance ranges of 1 km each. For New York, the closest source and destination we pick are only 1 km apart while the furthest are 16 km. For California, the closest source and destination we pick are only 5 km apart while the furthest are 80 km. For each range, we randomly pick 60 sets of sources and destinations that are within the geographical distance range. We treat them as the current location and the furthest point of interest of a querying vehicle respectively. We then consider the worst case that all points between the current location and the furthest point of interest are of the driver's interest. The types of information we consider are average vehicle speed and general road condition (e.g. accident, traffic jam). Without loss of generality, we assume that the vehicle requests for a credential or sends out its query once it enters an RSU's range (upon hearing its beacon broadcasts). Since a vehicle can wait for a random delay or travel for a random distance after obtaining a credential before sending out its query, we define the processing time as the period from when the vehicle sends out its query to when it finishes verifying the information provided by all RSUs in the reply message. This processing time is then normalized by dividing it by the duration that the vehicle is in the range of the RSU to which it sends its query. The data from all the 60 sets are then averaged to obtain a data point as shown in Figure 4 below. Note also that we represent a range using its class mark.

8.2. Simulation results

Fig. 4 shows the results for New York city. We can see that as the geographical distance increases, the processing time increases. When the source and the destination nodes are further away (i.e. a vehicle wants to gather information about a point of interest which is further away), more RSU hops are involved. This not only leads to more RSU signing operations but also more pairing operations at the vehicles in the verification phase. Nevetheless, among all geographical distance ranges, the processing time is at most 1.6 % of the duration that the vehicle stays in the querying RSU's range. Thus there must be sufficient time for the vehicle to finish its query and to verify the returning information.

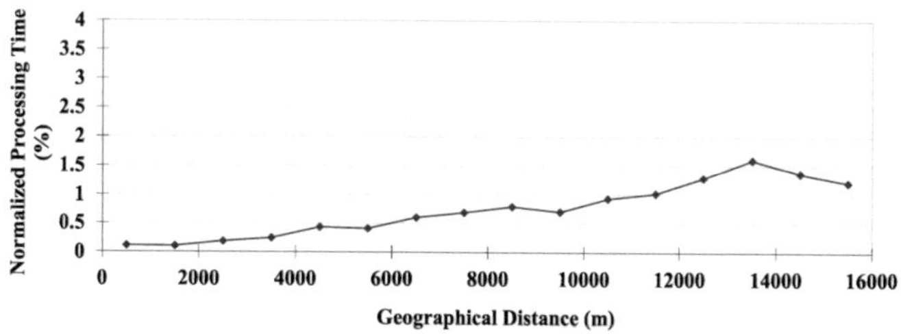

Figure 4. Normalized Processing Time vs. Geographical Distance (New York)

Fig. 5 shows the results for California city. We can see again that as the geographical distance increases, the processing time increases. When the source and the destination nodes are further away (i.e. a vehicle wants to gather information about a point of interest which is further away), more RSU hops are involved. This not only leads to more RSU signing operations but also more pairing operations at the vehicles in the verification phase. Among all geographical distance ranges, the processing time is at most 3.7 % of the duration that the vehicle stays in the querying RSU's range. This value is a little bit greater than that for New York city due to larger geographical distances. Anyway, there must be sufficient time for the vehicle to finish its query and to verify the returning information.

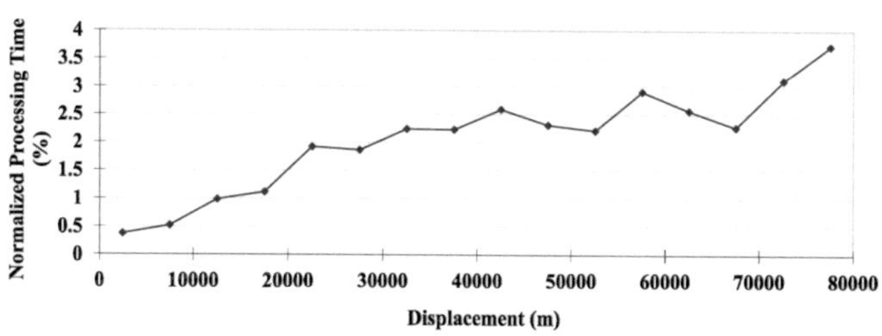

Figure 5. Normalized Processing Time vs. Geographical Distance (California)

9. Related work

A similar scheme of real-time information gathering using VANET is proposed in a recent work [17]. However, there are a number of differences between their scheme and ours. First, their scheme is a small scale navigation scheme which covers a carpark while ours is large scale to cover the whole city and beyond. Second, in their scheme a carpark is monitored by three RSUs which centrally take up the roles of determining a vehicle's location, searching for a vacant parking space and providing navigation service to guide the vehicle to go from the carpark entrance to the selected parking space. That is, all information are provided by the three RSUs. In our scheme, the road system in the city is monitored by a large number of RSUs which take up the information gathering task in a distributed manner. Third, in terms of security functions, their scheme assumes RSUs to be fully trusted. This makes sense since the three RSUs are installed indoor and can be monitored by security guards from time to time. However, such an assumption is no longer valid in our outdoor setting. It is impossible to have security guards monitoring all RSUs across the city. Thus, unlike their scheme, authentication of RSUs becomes a vital component in ours. Fourth, our scheme allows one's identity and query to be delinked. This feature is only interesting for wide area information gathering like ours. Thus, the scheme provided in [17] cannot be used to solve the information gathering problem discussed in this chapter.

Other recent efforts related to the security issues in VANET include [8, 9, 18–21]. In [8], a batch verification scheme was proposed for an RSU to verify a large number of signatures at the same time using only three pairing operations. In [18], an RSU-aided inter-vehicle communications scheme was proposed. A vehicle relies on an RSU to verify the signature of another vehicle. In [19], group communications in VANETs are considered and a group key update protocol was proposed. In [9], some security and privacy-enhancing communications schemes were proposed. Of particular interest, a group communications protocol was defined. After a simple handshaking with any RSU, a group of known vehicles can verify the signature of each other without any further support from RSUs. A common group secret is also developed for secure communications among group members. [20] and [21] also target at driver privacy preservation but instead of using pseudo identities, the concept of group signature is adopted. The signature of any vehicle can be verified by the same group key but the actual signer can only be traced by a trusted party. Though privacy can be preserved, group signature schemes are rather complicated and may not be practical.

10. Conclusions

In this chapter, we first highlighted the most significant security and privacy challenges in VANET protocol design. We then discussed how one should design security protocols for VANETs. In particular, we analyzed in details the advantages and disadvantages of hardware-based and software-based solutions. Next we proposed an information gathering scheme using VANETs. We utilized information collected by RSUs to provide drivers information about a set of points of interest that is out of sight in a distributed manner. Besides basic security features such as sender authentication and conditional identity privacy preserving. Our scheme adopts some security primitives in a non-trivial way to provide a number of additional security features: 1) With the idea of anonymous credential, no one including TA can link up a vehicle's query and its identity. 2) Queries and resulting information are protected from eavesdroppers. 3) Information provided by RSUs are

http://dx.doi.org/10.5772/53537

properly authenticated. We provided a security analysis and a simulation study to evaluate our scheme. In our simulation, we made use of the maps of New York (city road system) and California (countryside highway system) downloaded from the TIGER database. We found that the processing time is at most 1.6 % and 3.7 %, respectively, of the duration that the vehicle stays in the querying RSU's range in the two cities. Thus there must be sufficient time for the vehicle to finish its query and to verify the returning information.

Author details

T. W. Chim[1,2,⋆], S. M. Yiu[1], Lucas C. K. Hui[1] and Victor O. K. Li[2]

⋆ Address all correspondence to: twchim@cs.hku.hk; smyiu@cs.hku.hk; hui@cs.hku.hk; vli@eee.hku.hk

1 Department of Computer Science, The University of Hong Kong, Hong Kong
2 Department of Electrical and Electronic Engineering, The University of Hong Kong, Hong Kong

References

[1] H. Oh, C. Yae, D. Ahn, and H. Cho. 5.8 GHz DSRC Packet Communication System for ITS Services. In *Proceedings of the IEEE VTC '99*, pages 2223 – 2227, September 1999.

[2] J.P. Hubaux, S. Capkun, and J. Lui. The Security and Privacy of Smart Vehicles. *IEEE Security and Privacy Magazine, 2(3)*, pages 49 – 55, 2004.

[3] J. P. Hubaux M. Raya, P. Papadimitratos. Securing Vehicular Communications. *IEEE Wireless Communications, Vol. 13, Issue 5*, pages 8 – 15, October 2006.

[4] Y. Kim, A. Perrig, and G. Tsudik. Tree-Based Group Key Agreement. *ACM Transactions on Information Systems Security, 7(1)*, pages 60 – 96, 2004.

[5] Y. Jiang, M. Shi, X. Shen, and C. Lin. A Tree-Based Signature Scheme for VANETs. In *Proceedings of the IEEE GLOBECOM*, pages 1 – 5, 2008.

[6] C.K. Wong, M. Gouda, and S.S. Lam. Secure Group Communications using Key Graphs. In *Proceedings of the IEEE SIGCOMM*, pages 68 – 79, 1998.

[7] A. Menezes. An Introduction to Pairing-Based Cryptography. In *1991 Mathematics Subject Classification, Primary 94A60*, 1991.

[8] C. Zhang, R. Lu, X. Lin, P. H. Ho, and X. Shen. An Efficient Identity-based Batch Verification Scheme for Vehicular Sensor Networks. In *Proceedings of the IEEE INFOCOM '08*, pages 816 – 824, April 2008.

[9] T.W. Chim, S.M. Yiu, Lucas C.K. Hui, and Victor O.K. Li. SPECS: Secure and Privacy Enhancing Communications for VANET. *Elsevier Ad Hoc Networks, Vol. 9, Issue 2*, pages 189 – 203, March 2010.

[10] J. Jakubiak and Y. Koucheryavy. State of the Art and Research Challenges for VANETs. In *Proceedings of the IEEE CCNC*, pages 912 – 916, 2008.

[11] S.B. Lee, G. Pan, J.S. Park, M. Gerla, and S. Lu. Secure Incentives for Commercial Ad Dissemination in Vehicular Networks. In *Proceedings of the ACM MobiHoc*, pages 150 – 159, 2007.

[12] T.W. Chim, S.M. Yiu, Lucas C.K. Hui, and Victor O.K. Li. MLAS: Multiple Level Authentication Scheme for VANETs. *Elsevier Ad Hoc Networks Journal, Vol. 10, Issue 7*, September 2012.

[13] D. Boneh, B. Lynn, and H. Shacham. Short Signatures from the Weil Pairing. In *Proceedings of Asiacrypt '01*, pages 514 – 532, 2001.

[14] J. Baek, B. Lee, and K. Kim. Secure Length-Saving ElGamal Encryption under the Computational Diffie-Hellman Assumption. *Lecture Notes in Computer Science - Information Security and Privacy, Vol. 1841*, pages 49 – 58, 2000.

[15] Topologically Integrated Geographic Encoding and Referencing system (TIGER), 2009. http://www.census.gov/geo/www/tiger/.

[16] Google Map. http://maps.google.com.

[17] R. Lu, X. Lin, H. Zhu, and X. Shen. SPARK: A New VANET-based Smart Parking Scheme for Large Parking Lots. In *Proceedings of the IEEE INFOCOM '09*, pages 1413 – 1421, April 2009.

[18] C. Zhang, X. Lin, R. Lu, and P. H. Ho. RAISE: An Efficient RSU-aided Message Authentication Scheme in Vehicular Communication Networks. In *Proceedings of the IEEE ICC '08*, pages 1451 – 1457, May 2008.

[19] A. Wasef and X. Shen. PPGCV: Privacy Preserving Group Communications Protocol for Vehicular Ad Hoc Networks. In *Proceedings of the IEEE ICC '08*, pages 1458 – 1463, May 2008.

[20] B. K. Chaurasia, S. Verma, and S. M. Bhasker. Message broadcast in VANETs using Group Signature. In *Proceedings of the IEEE WCSN '09*, pages 131 – 136, December 2008.

[21] A. Studer, E. Shi, F. Bai, and A. Perrig. TACKing Together Efficient Authentication, Revocation, and Privacy in VANETs. In *Proceedings of the IEEE SECON '09*, pages 1 – 9, June 2009.